GIS for Homeland Security

Gathering and analyzing intelligence

Protecting critical infrastructure

Responding to complex emergencies

Preparing for disease outbreaks and bioterrorism

Securing complex events

Mike Kataoka

ESRI PRESS

REDLANDS, CALIFORNIA

ESRI Press, 380 New York Street, Redlands, California 92373-8100

Copyright © 2007 ESRI

All rights reserved. First edition 2007
10 09 08 07   2 3 4 5 6 7 8 9 10

Printed in the United States of America

Library of Congress Cataloging-in-Publication Data
Kataoka, Mike, 1952-

GIS for homeland security / Mike Kataoka.—1st ed.

    p.   cm.

    ISBN 978-1-58948-155-8 (pbk. : alk. paper) 1. Emergency management–United States. 2. National security–United
States. 3. Terrorism–United States–Prevention. 4. Geographic information systems–United States. 5. Geographic
information systems–United States–Case studies. I. Title.

    HV551.3.K38 2007

    363.340285—dc22   2007026706

Ask for ESRI Press titles at your local bookstore or order by calling 1-800-447-9778. You can also shop online at
www.esri.com/esripress. Outside the United States, contact your local ESRI distributor.

ESRI Press titles are distributed to the trade by the following:

In North America:
Ingram Publisher Services
Toll-free telephone: (800) 648-3104
Toll-free fax: (800) 838-1149
E-mail: customerservice@ingrampublisherservices.com

In the United Kingdom, Europe, and the Middle East:
Transatlantic Publishers Group Ltd.
Telephone: 44 20 7373 2515
Fax: 44 20 7244 1018
E-mail: richard@tpgltd.co.uk

Cover and interior design by Savitri Brant
Photograph of U.S. flag, PNC/BrandX Pictures/Jupiterimages.
Photograph of U.S. map detail, Ryan McVay/Photodisc/Getty Images.
Photograph of man in fire protection suit, D. Falconer/PhotoLink/PhotoDisc/Getty Images.
Photograph of children, Comstock/Comstock Images/Early Education/Jupiterimages.
Photograph of Jefferson Memorial, Kent Knudson/PhotoLink/PhotoDisc/Getty Images.
Photographs of the U.S. capitol building, PNC/BrandX Picures/Jupiterimages.

Contents

Preface

"Homeland security" entered the national consciousness just a few years ago, but the planning, preparation, analysis, and action implicit in those words have long been the function of government and geography. Geospatial technology is a proven, essential tool for mapping locations, visualizing dynamic conditions, and making informed decisions during emergencies. In a homeland security context, geographic information systems (GIS) answer the key questions, starting with who and what are at risk.

Securing the homeland from terrorist attacks and natural disasters involves multiple jurisdictions and vast quantities of diverse data. When minutes matter, the integrating technology of GIS sets the foundation for strategic response to protect lives, infrastructure, and resources. The GIS framework also helps assess vulnerabilities and prepare for homeland threats. Sharing geographic information across multiple jurisdictions requires heightened levels of cooperation, strong alliances, political foresight, and technological innovation.

The September 11, 2001, terrorist attacks and Hurricane Katrina, among other recent catastrophic events, have underscored the urgency for consistent, effective, and cost-efficient ways to meet the homeland security mandate. When the U.S. Department of Homeland Security (DHS) was created in 2002, its primary function was to bring together disparate entities committed to deterring terrorist attacks and protecting against and responding to threats and hazards. GIS immediately played a key role in the DHS mission areas, from prevention to recovery. The department reaffirmed its commitment to GIS in 2006 when it signed an enterprise license agreement with ESRI. This agreement enabled DHS to expand enterprise GIS technology for sharing information and analysis across the department.

Numerous jurisdictions that face homeland security challenges have developed GIS-based solutions worth emulating. Several are presented as case studies in *GIS for Homeland Security*. With a minimum of technical language, this book concisely illustrates how GIS functions behind the scenes in the real world, including events as high profile as the Super Bowl and Olympic Games. It describes GIS at work on the front lines of wildfires and explosions, and on computer networks that monitor terrorism, crime, and the spread of disease.

GIS for Homeland Security also singles out visionaries who have embraced spatial technology in the field. These often unsung heroes have applied creative thinking, determination, and old-fashioned legwork to the evolving technology, resulting in model approaches to homeland threats.

Many can benefit from this book, from top-level decision makers and first responders on the ground, to experienced GIS users striving to push the technology to its limits, to those just getting acquainted with GIS.

Mike Kataoka
Redlands, California

Acknowledgments

GIS for homeland security requires collaboration and so did the publication of this book. Thanks to my colleagues at ESRI, especially Russ Johnson, Paul Christin, Jennifer Schottke, Jeff Sopel, Tom Patterson, Kasey Quayle, and Jonathan Fisk who contributed greatly to the cause of presenting relevant and readable material.

I'm indebted to the many dedicated GIS professionals across the country who lent their expertise and valuable time to this project by reviewing text, offering suggestions, and providing images. Chief among them were Matthew Felton, Towson University; Soheila Ajabshir, Miami-Dade County; Rusty Wynn and Steve Wood, City of Pleasanton; Eric Olson, PureTech Systems; Teresa Woods, South Carolina Law Enforcement Division; T. James Fries, PlanGraphics; Scottee Cantrell, Duke University; Chris Goranson, New York City Department of Health and Mental Hygiene; Lauren McLane, Federal Emergency Management Agency; Johanna Meyer, Massachusetts Emergency Management Agency; and Dale Blasi and Mary Lou Peter-Blecha, Kansas State University.

I'm fortunate to work with a team of creative professionals at ESRI Press led by Judy Hawkins, Dave Boyles, Michael Hyatt, and Kathleen Morgan. I'm especially grateful to Dave for fine-tuning the manuscript. This project reflects the considerable talents of designers Savitri Brant and Jennifer Galloway, graphic artist Jay Loteria, copy editor Tiffany Wilkerson, administrative assistant Kelley Heider, and print production supervisor Cliff Crabbe. My work also reflects the loving support of my wife, Jeanne, and my children, Kris and Lisa.

Jack and Laura Dangermond deserve special thanks for sustaining an environment here at ESRI that encourages innovative thinking and technological solutions to global challenges.

Introduction

Geographic information systems (GIS) and homeland security is as much about breaking down barriers as fortifying defenses. Terrorist attacks and natural disasters have validated GIS as the technology of choice when seconds count and lives and property are at risk. Crisis situations, however, have also exposed significant gaps in coordination among bureaucracies slow to change.

In 2005, coinciding with the fourth anniversary of the September 11, 2001, attacks, the GIS community convened in Denver for a homeland security summit. Participants hailed the state of the technology and praised GIS professionals who were aiding Hurricane Katrina victims that very week. But back-patting was tempered with hand-wringing as some voiced frustration with the political, financial, and logistical impediments—as manifested in the Gulf Coast—that have kept GIS from reaching its full potential as a collaborative tool. *GIS for Homeland Security* focuses on many positive trends in the face of considerable challenges.

This book chronicles the accomplishments of creative people and progressive programs embracing GIS technology to assess danger, coordinate response, and facilitate recovery. Case studies from coast to coast illustrate how government and private entities have developed model solutions to real threats. Also, *GIS for Homeland Security* profiles visionaries who have applied the technology in the field—including the national stage—with impressive results.

Mapping and mitigation

GIS technology blends geographic and descriptive information to produce digital maps with multiple layers. By linking maps to databases, GIS helps users understand location and visualize data so that they can act decisively and effectively while adapting to shifting trends. People use GIS to analyze the vulnerability of buildings, bridges, power plants, ports, and other critical spatial features. GIS helps them devise mitigation strategies, whether by limiting access, posting armed guards, toughening targets, or devising alternate routes. Some GIS action plans can be tailored to facilities that are at risk at certain times, such as a football stadium, concert venue, or shopping mall.

During a catastrophic event, data sharing and accessibility are crucial. After the terrorist-caused destruction of the World Trade Center in New York City, vital geospatial data existed but was stored in scattered locations with no central administration. Those on the scene first had to create an integrated GIS before they could use the data for analysis, response, and recovery. They scoured city and private records for infrastructure data, imported it into a GIS, and produced valuable maps, including some used to deploy emergency vehicles to the highest priority areas.

A regional approach

Many took the lessons of September 11 to heart and intensified their commitment to collect GIS data for homeland security applications. They have developed regional, coordinated GIS programs that promote the flow of intelligence, protect critical infrastructure, improve emergency response, secure high-profile events, and prevent the spread of toxic chemicals and disease. Five chapters in *GIS for Homeland Security* explore these key topical areas, each with three case studies detailing actual events and operations. A final chapter looks at the evolving role of GIS, particularly in federal initiatives aimed at integrating geographic data.

The data-sharing concept has been hindered, in part, by the impulse among governments and companies to claim a proprietary right to their valuable assets, including geographic information. For years, public and private entities have established GIS databases and invested considerable money and time into software, equipment, training, and operating procedures.

Most federal agencies, including the U.S. Geological Survey, the Environmental Protection Agency, and the justice, agriculture, and homeland security departments, maintain extensive geodatabases, as do governments at the state and local level and the private sector. But much of this abundant data is isolated within departments and organizations, many of which are convinced that relaxing secrecy would erode security. Jurisdictional policy differences over how data is used, legal issues such as intellectual property rights, and concerns about competition, security, and privacy all frustrate data sharing.

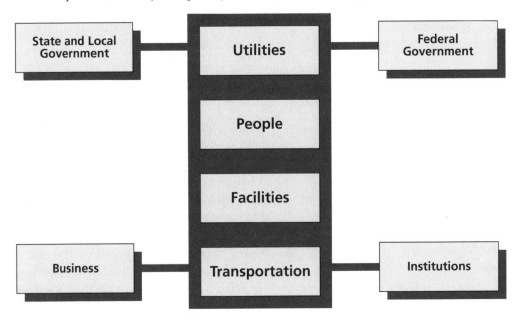

Government and the private sector coordinate policies, resource allocation, and partnership strategies to fulfill the homeland security mission. Within these organizations are divisions and departments that manage data about people, places, and things, all critical to plan for and respond to emergencies.

Source: ESRI.

The federal government is pushing initiatives to expand the role of geographic technology in homeland security—with data sharing a top priority—so that the nation's digital data is instantly accessible, accurate, timely, and compatible. That's a tall order that GIS professionals agree is best accomplished from the grassroots up. And geospatial data cannot be fully exploited in emergency situations unless police, firefighters, and paramedics at the scene can navigate a GIS as easily as a two-way radio.

The U.S. Department of Homeland Security's National Operations Center collects and fuses information from more than thirty-five federal, state, territorial, tribal, local, and private-sector agencies, constantly providing real-time situational awareness and monitoring.

Source: U.S. Department of Homeland Security.

Homeland security GIS is an outgrowth of spatial technology applied to wildfires, earthquakes, floods, and other natural disasters. The consequences of a malicious attack, much like an act of nature, can be mitigated through planning and preparedness. Homeland security, above all, means preventing catastrophe. That works most efficiently when diverse parties pool vital information about hazardous materials, key transportation routes, and vulnerable locations.

Through GIS, people with different functions within the same organization—such as the fire chief, public utilities manager, and transportation director—come together to develop a common homeland security vision. The data-gathering process also fosters relationships among neighboring public officials who otherwise would not interact except in an emergency.

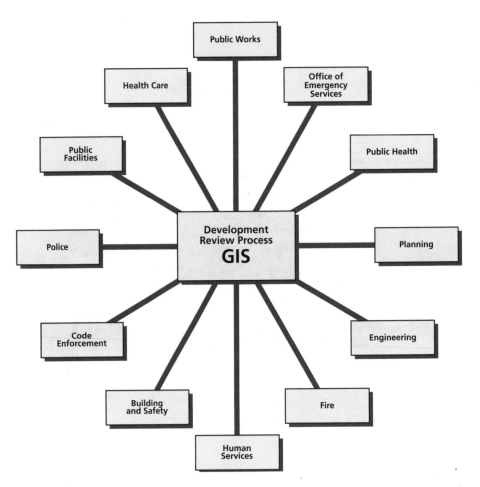

GIS is the core technology that helps manage geographic data from various governmental operations that is crucial in an emergency or homeland security situation.

Source: ESRI.

Grants based on regional planning

In January 2006, U.S. Homeland Security Secretary Michael Chertoff emphasized the importance of regional cooperation as the basis for awarding $765 million in Urban Areas Security Initiative (UASI) antiterrorism grants to regions: "The bottom line is this: When terrorists plan an attack, the attack is not carefully delineated within the lines of political jurisdictions," Chertoff said.[1] The same holds true, he said, for hurricanes and other natural disasters.

Regional organizations have found that shared GIS data, quickly and easily accessible among multiple agencies, quells the inevitable chaos associated with catastrophic events.

1. Remarks by Homeland Security Secretary Michael Chertoff at a press conference on the fiscal year 2006 Urban Areas Security Initiative Grants, January 3, 2006.

Secretary of Homeland
Security Michael Chertoff

Source: U.S. Department of
Homeland Security.

They realize that real-time functionality, or as close to it as possible, is essential to visualize patterns, map locations, and grasp the spatial context of volatile situations. Regional GIS programs succeed when people who share responsibility for homeland security planning and emergency response develop close interagency ties. A crisis situation is too late to get acquainted.

Collaborating agencies bonded by GIS technology can identify shared problems and pinpoint where combining resources makes sense to protect critical assets and carry out other homeland security functions. This regional approach can be accomplished through a central shared data repository or distributed networks. Many GIS homeland security applications are Web-based and controlled by multijurisdictional data-sharing agreements with password-protected access to sensitive data.

As demonstrated in the aftermath of Hurricanes Katrina and Rita in 2005, geographic information is extremely valuable when responding to and recovering from natural disasters. Dozens of GIS volunteers helped officials coordinate rescues and evacuations, monitor weather, assess damage, and facilitate recovery. GIS, in conjunction with Global Positioning Systems (GPS) and other technology, enables incident commanders to get the upper hand on a fire, flood, chemical cloud, or similar event by anticipating its spread and direction and deploying personnel accordingly.

Modeling the consequences of an attack or natural disaster helps policy makers justify spending for preparedness, response, and mitigation. GIS has the capacity to integrate geographic data with threat intelligence. By mapping the proximity of potential threats to critical infrastructure, decision makers can ward off an attack or disrupt terrorist plans.

GIS is involved in virtually all phases of homeland security and is especially effective when multiple entities work together to preserve the country's security, health and safety, economic strength, and way of life. *GIS for Homeland Security* is a glimpse at a monumental movement in geospatial technology that will, and must, continue to build.

1
Gathering and analyzing intelligence

Homeland security intelligence is more about bits and bytes than cloaks and daggers these days, and GIS is the core technology for collecting, analyzing, and visualizing spatial data. Tom Ridge, the first U.S. secretary of Homeland Security, pictured an ideal counterterrorism computer network linking federal, state, and local intelligence and law enforcement agencies so they could instantaneously share threat reports, investigative leads, and potential evidence.

"In this new post-9-11 era, a new philosophy is required—a philosophy of shared responsibility, shared leadership and shared accountability," Ridge said in 2004 when he announced the creation of a national intelligence network. "The federal government cannot micromanage the protection of America."[1]

1. Spencer S. Hsu, "Anti-Terrorism Network Launched. System Allows Agencies Across Country to Share Data Instantaneously." *The Washington Post*, February 25, 2004, B01.

Computer technology, principally GIS, helps government officials monitor crimes and suspicious activity suggesting terrorism. Collecting and fusing spatial data from a broad array of sources is the backbone of contemporary antiterrorism intelligence. Data sources consist of official law enforcement intelligence along with general information from other government agencies and the private sector. Data fusion supports risk-based, information-driven programs of prevention and response to emerging or immediate threats, be they plots by humans or forces of nature.

Surveillance and response

In the areas of surveillance and detection, GIS can help law enforcement and intelligence officials grasp risky situations by assessing time-and-space relationships, recognizing patterns, and correlating seemingly disconnected events. Digital mapping coordinates surveillance and detection while the analytical capabilities of GIS help decision makers take defensive or preemptive action. A geographic analysis can track the movements of suspicious people, mobilize resources efficiently to prevent a terrorist event or minimize consequences, and determine an area's vulnerability to bioterrorism.

Increasingly sophisticated information-gathering systems, from covert sensors to satellite communications interceptors, generate huge volumes of data much like pieces of a jigsaw puzzle. The integrating power of GIS can create a vivid panorama. When threat data is matched with geographic data, GIS can document the convergence of location, vulnerability, high risk, suspects, and events. Statewide organizations such as the California Anti-Terrorism Information Center, established shortly after the September 11 attacks, and the Arizona Counter Terrorism Information Center, formed in 2004, have emerged as models for multifaceted intelligence processing powered by GIS.

Secure data sharing

The premise of the Department of Homeland Security's intelligence-sharing policy is that a widespread Internet-based network must be easily accessible through official channels yet secure enough to prevent sensitive information from getting into the wrong hands. Sharing geographic data in an open society raised concerns that potential terrorists could exploit this information. But a 2004 study by RAND, the California think tank, concluded that there was no need to impose broad restrictions on digital geographic data. The study found that the vast majority of federal datasets are of little use to an attacker or are available from so many other nongovernment sources that limiting access would be pointless. The study further found that data sharing is good for the economy as well as homeland security.

Intelligence gathering and analysis pays dividends to state and local governments beyond terrorism awareness. The process strengthens their ability to identify and forecast emerging crime, public health, and quality-of-life trends; supports targeted law enforcement and other problem-solving activities customized to local communities; and improves emergency and nonemergency services.

As outlined by the U.S. Homeland Security Advisory Council in 2005, effective intelligence fusion requires common terminology, definitions, and lexicon; a shared understanding of the threat environment and indicators of an emerging threat; coordination with federal intelligence and law enforcement entities on information-gathering protocols; and clear delineation of roles, responsibilities, and requirements at each level involved in the fusion process.

The president and Congress called for establishing an "information sharing environment" in which data fusion and collaboration activities within the federal government and among federal, state, local, tribal, and private entities carry out the homeland security mission.

Such an environment not only involves law enforcement, and state and regional intelligence centers, but also a general public educated on what suspicious activity or circumstances to look for and what to do with that information.

CASE STUDY

Emergency Management Mapping Application (EMMA)

A densely populated region containing the seat of federal government, financial centers, and numerous high-tech companies requires exceptional safeguarding from terrorist acts and natural disasters. Maryland and the National Capital Region have put their trust in the Emergency Management Mapping Application (EMMA). The data fusion software provides the tools to pinpoint an incident, generate a location report with a visual display, analyze an affected area, and coordinate resources. EMMA fosters communication among disparate agencies and assists emergency crews in the field by creating a common operating picture.

EMMA, shorthand for the Emergency Management Mapping Application data fusion software, is a vital tool for homeland security officials responsible for protecting the National Capital Region.

Comstock/Comstock Images: Government & Social Issues/ Jupiterimages.

The Web-based GIS enterprise system was developed at Towson University. It is built on a Java framework and integrates various ESRI-based mapping capabilities that follow industry standards. EMMA is installed at the Maryland State Emergency Operations Center and allows users to access various databases through interactive maps. EMMA's interoperability benefits dozens of cities, counties, police departments, and transportation agencies in the region. That technological edge has resulted in coordinated action across all levels of government.

Towson creates EMMA

The Center for Geographic Information Sciences (CGIS) at Towson University near Baltimore, Maryland, created EMMA in 2003 for the Maryland Emergency Management Agency to satisfy the need for mapping technologies that save time and lives. That year, Hurricane Isabel unleashed fierce winds and storm surge flooding. Within minutes, EMMA determined where to place sandbags at the Frederick County reservoir, a task that would have taken hours in the field.

In 2005, the U.S. Department of Homeland Security's Information Technology Evaluation Program funded a pilot effort called the Maryland Emergency Geographic Information Network (MEGIN). Led by Towson University, the pilot established technology for securely sharing data across organizational, jurisdictional, and disciplinary boundaries. MEGIN is Maryland's vision for a secure, coordinated information portal for the emergency management community, whose goal is to get the right data to the right person at the right time. MEGIN comprises ESRI's Portal Toolkit with an added measure of security at multiple levels, as well as a process for data owners to securely share map services with regional partners.

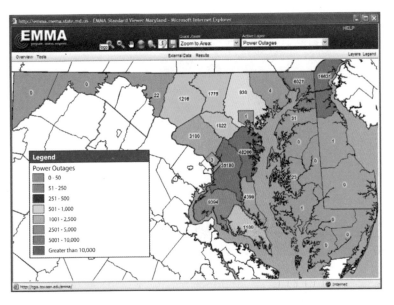

EMMA integrates real-time power outage information across Maryland. This map was produced during a 2007 ice storm.

Sources: Matthew Felton, Towson University; Maryland Emergency Management Agency.

EMMA is fully interoperable with other GIS software and with incident management software. Its modular system architecture allows for flexibility and expandability, so as more maps are needed, more servers can be used for both database and application scalability. EMMA incorporates the Open GIS Consortium (OGC) Web Map Services (WMS) standard, which is an XML- and URL-based schema for providing and retrieving map services across multiple platforms.

In an emergency situation, EMMA pulls data from various agencies and weaves it into a single interactive map. First responders have access to crucial resources, such as nearest police and fire departments, vulnerable areas, and local geography. EMMA's real-time capabilities include determining a particular hospital's available beds, calling up current weather conditions, and pinpointing the location of individual patrol cars. When hazardous materials spill on the highway, EMMA tells users the direction and speed of the wind, which schools are in harm's way, and what roads need to be closed.

The reach of EMMA

EMMA is tied to several federal and state GIS initiatives, including the U.S. Geological Survey's National Map project and Geospatial One-Stop, the Web-based data exchange program. EMMA plays an important role in developing mitigation plans for critical infrastructure, modeling emergency events, and response plans. CGIS has worked closely with the State Emergency Operations Center (SEOC) to put technologies and processes in place that significantly increase Maryland's preparedness for large-scale natural or terrorist disasters.

In 2005, Maryland relief officials were deployed to the Gulf Coast region devastated by Hurricane Katrina. Response teams found themselves in unfamiliar territory and with limited communications. EMMA proved to be the equalizer, coordinating the operation with

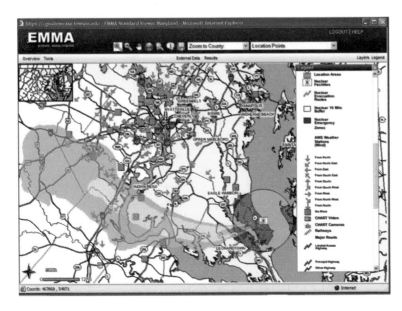

EMMA incorporates real-time weather and transportation data with a nuclear plume simulation in the National Capital region.

Sources: Matthew Felton, Towson University; AWS Convergence Technologies; Maryland Department of Transportation; National Oceanic and Atmospheric Administration.

dynamic maps. Dubbed "EMMA Katrina," the operation coordinated data for a broad area of the Gulf Region, and its integration with WebEOC allowed for the location of response teams to be tracked practically in real time.

Building on the success of Maryland's statewide implementation of EMMA, local implementations were rolled out to individual county jurisdictions in Maryland. In addition, several other states as well as local jurisdictions within the National Capital Region looked toward EMMA as a GIS solution to their emergency operations needs.

During the summer of 2006, a prototype implementation of EMMA dubbed "NCR Map" was deployed to demonstrate a regional implementation of EMMA in support of the GIS committee of the Metropolitan Washington Council of Governments (MW COG). Shortly after this prototype was deployed, emergency management officials from Washington, D.C., used EMMA to understand the risk posed by a leaking dam several miles north of their jurisdiction in Rockville, Maryland. Rather than requesting this data and importing into their GIS, Washington, D.C., officials simply launched EMMA and focused on the area at risk.

Reusability

EMMA's latest release, version 2.0, focuses on increased security and interoperability, additional tools, and an expanded database. Future versions of EMMA will be designed to access information such as geospatial metadata through MEGIN by providing a search engine.

In February 2006, both EMMA and MEGIN were evaluated by independent consultants from the National Capital Region who assessed the technical reusability of various programs in the region. As part of this process, EMMA and MEGIN technology was measured against the federal government's Technical Reference Model (TRM), a part of the E-GOV initiative that provides a foundation to categorize the standards, specifications, and technologies to support the construction, delivery, and exchange of business and application components. EMMA and MEGIN technology received very high marks in this evaluation, which verified its reusability for a broad range of applications.

CASE STUDY

South Carolina Information Exchange (SCIEx)

Geospatial technology has the capacity to transform accumulated information into valuable intelligence for enforcing the law and protecting the homeland. That's exactly what has happened in South Carolina where a grassroots cooperative among sheriffs and police chiefs in 2001 blossomed into a statewide model for intelligence fusion.

After the September 11 attacks, South Carolina's state and local law enforcement community saw the need for a system to gather information from diverse, statewide resources and make the data available to the intelligence process. To develop actionable intelligence, analysts must have data in useable formats from as wide a universe of resources as practical and pertinent. This capability was largely missing or fractured around South Carolina's agencies and local jurisdictions.

A proactive approach

The South Carolina Law Enforcement Division (SLED) set out to develop an information technology system to support the intelligence process. Its mission would be to take a proactive approach to collecting, displaying, analyzing, and acting on intelligence in order to detect and prevent terrorist acts and precursor crimes such as money laundering, identity theft, narcotics, and smuggling.

In March 2005, the South Carolina Information Exchange (SCIEx) was conceptualized and initiated. It would consist of the SCIEx Intelligence Fusion Center and the SCIEx Data Warehouse project. An open-source model for sharing law enforcement records management system (RMS) incident information in Charleston—in which six agencies in a four-county area shared incident data electronically across jurisdictional boundaries—was adopted as the core technology to develop and expand statewide.

The task was significant, in that South Carolina has over 275 law enforcement jurisdictions and over twenty RMS application vendors with active accounts in the state. SLED contracted with the developers of the model, the National Law Enforcement Corrections and Technology Center-Southeast (NLECTC-SE) and Scientific Research Corporation (SRC) to continue to expand and grow the systems capabilities and footprint. The system currently has more than 150 agencies replicating incident data to the warehouse and using the system in the conduct of investigations.

GIS joins the picture

Early in the development of the SCIEx project it was determined that GIS would be invaluable for the display and analysis of the raw incident data. SLED turned to ESRI and its business partner the Omega Group, a San Diego–based GIS specialist for law enforcement, and they teamed with SRC to implement the geospatial component of SCIEx. The technical team first geocoded all crime data stored in the SCIEx warehouse and designed a process for geocoding new crimes entered into the system. The Omega Group's CrimeView analytical software then examined the geocoded data for trend analysis, querying, and reporting.

A common operational picture (COP) situation map was developed to display incidents and data layers on the fusion center video wall. Near-real-time incidents were displayed against layers of critical infrastructure, potential targets, areas of interest or activity, and other datasets collected from around the state. Because the system was open and scaleable, SCIEx was able to integrate the state's sex offender dataset into the central data warehouse and worked toward adding many more datasets.

South Carolina is one of thirteen states committed to a homeland security initiative called Southern Shield. Relying on GIS, the member states exchange best practices, share terrorism-related intelligence, and monitor regional terrorism threats. The Global Justice XML data reference model, sponsored by the U.S. Department of Justice, has facilitated the interstate and federal collaboration.

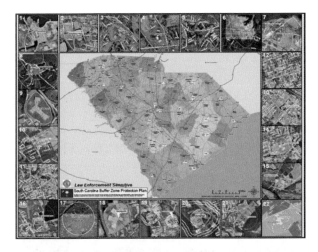

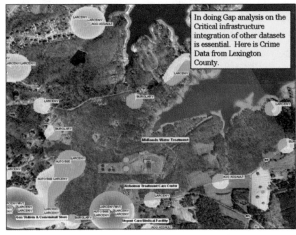

South Carolina law enforcement relies on GIS to coordinate intelligence and data on crime and infrastructure to provide a clear picture of threats and potential targets.

Source: South Carolina Law Enforcement Division.

SCIEx would be expanded to include data from corrections and jail entities, computer-aided dispatch systems, the court system, and a host of other state agency data. The SCIEx Fusion Center also planned to introduce Pictometry technology—oblique aerial photography integrated with GIS—for critical infrastructure protection and response, and in support of operations like natural disaster recovery and major event security.

Captain Teresa Woods of the South Carolina Law Enforcement Division was principal contributor to this case study.

CASE STUDY

Arizona Counter Terrorism Information Center (ACTIC)

Arizona's counterterrorism network combines high-tech processes with grassroots participation to secure a vulnerable corner of America's homeland. The Arizona Counter Terrorism Information Center (ACTIC) is the state's analysis hub for crime- and terrorism-related intelligence and is staffed by two hundred representatives of local, state, and federal law enforcement agencies. Based in Phoenix, ACTIC has investigated security breaches at Sky Harbor International Airport, an attempt to manufacture the biological toxin ricin, and suspicious international border crossings. A $5.3 million Department of Homeland Security grant in 2004 established the center, hailed as a model for cooperative interagency law enforcement.

A dynamic data fusion system is ACTIC's most potent weapon in its mission to detect, prevent, and respond to terrorism and other critical events. The system, powered by ArcGIS software and MetaCarta's Geographic Text Search (GTS), combines structured intelligence and spatial data with such unstructured content as e-mail, Web pages, and news articles, enriching intelligence for field agents.

Tapping unstructured information sources

GTS automatically extracts geographic references from unstructured text and plots documents on a map. Analysts can search text archives using keywords and geographic extents as filters. Without this capability, analysts would spend hours pouring over documents relevant to the area in question. ACTIC agents routinely use GTS technology to check public sources of unstructured information, such as American and Mexican newspaper Web pages, when gathering intelligence about a specific location. So, when law enforcement responds to an incident, officers instantly know what has happened in the

The Arizona Counter Terrorism Information Center gathers intelligence to protect the state, including its vast reaches of hostile desert.

Photo 24/BrandX: National Parks/ Jupiterimages.

Surveillance along the Arizona border includes use of unmanned aerial vehicles equipped with electro-optic sensors and communications equipment to provide around-the-clock images to Border Patrol agents.

Source: U.S. Department of Homeland Security.

neighborhood that has made news or drawn the attention of other agencies. MetaCarta officials and law enforcement authorities estimate that 80 percent of intelligence relevant to law enforcement is unstructured content.

Similar to the Neighborhood Watch concept, ACTIC encourages citizens to be vigilant and report any suspicious activity. As terrorists focus on buses, trains, public gathering spots, and other "soft targets," ACTIC relies on ordinary citizens to step up as the eyes and ears of the community.

Sharing the data

Calls to the center are assigned to an agent who logs the report into a database and refers it to the appropriate department, be it the bomb squad or counterterrorism unit, to determine if the threat is legitimate. From that point on, all public safety agencies from the federal government to the local fire departments have access to the same data. Before ACTIC, the data information systems of various law enforcement entities were not linked, making it difficult to track suspicious trends or recognize a pending terrorist threat.

Government intelligence reports consider Arizona's 370-mile international border to be vulnerable to terrorist penetration, so that stretch of desert is patrolled by some 2,400 agents and civilian volunteers. Law enforcement officials say that while al-Qaeda appears to favor safer, easier routes into the United States over Arizona's harsh southern boundary, the threat remains real. Border control has been beefed up dramatically since September 11, with more labor and technology focusing on antiterrorism. Agents are trained to spot non-Mexicans and turn suspicious border crossers over to the FBI for questioning.

ACTIC also assists the private sector's contribution to homeland security. The center created an information-sharing and training program for about 19,000 security officers attached to 201 private companies throughout Arizona. Homeland defense experts say that private security companies protect three-fourths of the nation's most likely targets for terrorism. Arizona relies on private security to patrol dam sites and airport perimeters, protect the state's nuclear power plant, and guard banks. In 2002, state officials toughened standards for private security guards and boosted their role as community sentinels who have access to shared law enforcement intelligence and are plugged into the alert system when threats arise.

References

Arizona Counter Terrorism Information Center Web site. http://www.nga.org/cda/files/ 0405BioterrorismPhelps.ppt.

Baker, John C., et. al. 2004. *Mapping the risks: Assessing the homeland security implications of publicly available geospatial information.* Santa Monica, Calif.: RAND Corporation.

Coleman, Kevin. 2003. GIS, information technology, and biotech take center stage in supporting homeland security. *Directions Magazine,* April 11. http://www.locationintelligence.net/ articles/345.html.

E-Gov. Federal Enterprise Architecture. http://www.whitehouse.gov/omb/egov/a-6-trm.html.

Felton, Matthew, and John Morgan. 2004. Emergency management mapping application: Integrating data online for emergency management. Paper presented at the twenty-fifth annual ESRI International User Conference, August 9–13, San Diego, California.

GIS for Emergency Management in Maryland. 2004. EMMA: Emergency Management Mapping Application Towson University Center for Geographic Information Services Quarterly Newsletter, spring. http://cgis.towson.edu/newsletter/newsletter_spring_04_emma.htm.

Hensley, J. J. 2005. Phoenix center a hub for coordinating terrorism data. *The Arizona Republic,* July 9. http://www.azcentral.com/arizonarepublic/local/articles/0709counterterror.html.

Justice Technology Information Network Web site. South Carolina Information Exchange PowerPoint presentation. www.justnet.org/nlectcse/download/knight_sciex_sccjis2006.ppt.

Knight, Coleman. Statewide information sharing in South Carolina. 2006. *The Police Chief* vol. 73, no. 4 (April). http://policechiefmagazine.org/magazine/index cfm?fuseaction=display_ arch&article_id=856&issue_id=42006.

Maryland Governor's Office of Homeland Security. 2005. *Public Safety Communications Interoperability in Maryland.* MD-IPT-RPT-R3CI. February 28.

Ridley, Randy, and John-Henry Gross. 2005. Preventing terrorism with geographic text searches. *ArcUser Online* April–June. http://gis.esri.com/library/userconf/proc04/docs/pap1430.pdf.

Ridley, Randy, and Mike Odell. 2005. Interview by Matteo Luccio, *GIS Monitor newsletter.* August 11. http://www.gismonitor.com/news/newsletter/archive/081105.php.

Romeo, Jim. 2005. Arizona gets a better view. *Emergency, Fire/Rescue & Police Magazine.* http://www.efpmagazine.com/Technology/MetaCarta.asp (accessed July 12, 2006).

SCRA Web site. Information is the key to security. http://www.scra.org/homeland_security.shtml.

South Carolina Law Enforcement Division. 2006. *2005–2006 Annual Accountability Report.*

Thorlin, Scott. Setting the table for information sharing: Would you please pass the intelligence? December 27, 2004, interview on FBI Web site. http://www.fbi.gov/page2/dec04/actic122704.htm.

Towson University Center for Geographic Information Services Web site. 2004. Fire? Flood? Rescue workers use TU's new CGIS mapping tool. *Tech Talk* March 29. wwwnew.towson.edu/techtalk/20040329,1_techTalkArticle.html.

———. EMMA: Prepare, assess, respond. http://cgis.towson.edu/downloads/EMMA_whitepaper.pdf.

U.S. Department of Homeland Security. Homeland Security Advisory Council. 2005. *Intelligence and Information Sharing Initiative: Homeland Security Intelligence & Information Fusion.* April 28.

Wagner, Dennis. 2004. Border no terror corridor—so far: Still, infiltration threat via Mexico called real. *The Arizona Republic,* August 22. http://www.azcentral.com/arizonarepublic/news/articles/0822borderterror23.html.

———. 2006. Private security guards play key roles post-9/11. *The Arizona Republic,* January 22. http://www.azcentral.com/arizonarepublic/news/articles/0122privatesecurity.html.

2
Protecting critical infrastructure

A prime function of homeland security is to identify and prioritize a community's vital buildings, systems, and resources, and protect that infrastructure from terrorist acts, natural disasters, or other emergencies. GIS eases the burden of this complex process.

Homeland Security Presidential Directive 7 (HSPD-7), issued on December 17, 2003, outlined the requirements for protecting critical infrastructure and key resources, including a nationwide plan relying in large part on geospatial technology. (Presidential directives establish national policies, priorities, and guidelines to strengthen U.S. homeland security.) As defined by the directive, critical infrastructure consists of high-profile buildings, such as the Pentagon and White House, along with the utilities, communications networks, transportation routes, military installations, hospitals, chemical plants, agriculture and food supply, financial systems, and other essentials of modern society.

In June 2006, the U.S. Department of Homeland Security (DHS) completed the National Infrastructure Protection Plan, which fine-tuned HSPD-7 and called for heightened coordination, integration, and synchronization of technical resources, including geographic data.

A tool for assessing and predicting

Mapping infrastructure data helps assess risks to specific locations and predicts the consequences of emergency events to people, property, the environment, and business. By exposing vulnerabilities of critical sites, GIS can help officials mitigate risk, whether by beefing up security, installing surveillance equipment, or restricting access. GIS integrates geographic data with threat intelligence to warn of "perfect storm" convergences of location, vulnerability, high risk, perpetrators, and opportunity. GIS can document the proximity of critical sites and display the extent of a buffer zone necessary to protect a facility and its neighbors.

The DHS's Buffer Zone Protection Plan (BZPP) funnels money from states to local governments for equipment to protect areas surrounding critical facilities. GIS technology, specifically ESRI software, helps officials meet the BZPP goals, which include defining the buffer zone, identifying specific threats and vulnerabilities and assigning a corresponding risk level, and recommending risk-reduction measures.

GIS helps define a buffer zone by mapping the proximity of targeted infrastructure to other critical sites, along with schools, hospitals, transportation routes, and the like. Officials aided by GIS can also predict what areas would be affected by a power outage, toxic plume, or other emergency, and share that information with affected agencies.

The CARVER method

Communities routinely prepare for natural disasters by identifying flood zones, timberland, earthquake faults, and other vulnerabilities. Protecting critical infrastructure broadens that vigilance. A widely used method for assessing the vulnerability of a specific asset is CARVER, an acronym derived from the first letters of each step:

- Criticality—Identify critical assets, single points of failure, or "choke points."
- Accessibility—Determine ease of access to critical assets.
- Recoverability—Compare time it would take to replace or restore a critical asset against maximum acceptable period of disruption.
- Vulnerability—Evaluate security system effectiveness against adversary capabilities.
- Effect—Consider scope and magnitude of adverse consequences that would result from malicious actions and responses to them.
- Recognizability—Evaluate likelihood that potential adversaries would recognize that an asset was critical.

Once vulnerabilities are identified and prioritized through this assessment process, GIS applications then can be applied to strengthen emergency response. Mapping functions can improve response time, deploy personnel more efficiently, and locate emergency resources more strategically. GIS modeling can calculate damage and casualties caused by an event and dictate planning and training procedures. Spatial information—access routes, utility shut-off values, hazardous storage areas—readily available through mobile technology, gives first-responders an important edge at the scene of an emergency.

Since January 2004, the DHS has encouraged critical infrastructure stakeholders from government and private organizations to address collaboration and data-sharing issues through local pilot projects. The idea is to promote an interactive, cooperative forum to determine data requirements, organizational process requirements, interoperability and enterprise architecture requirements, and technology requirements. The ultimate goal is a unified national geospatial framework for protecting critical infrastructure.

CASE STUDY

TriValley GIS

Lessons learned from the terrorist attacks on the urban East Coast reached communities across the continent, including Pleasanton, Livermore, Dublin, Danville, and San Ramon, collectively known as the TriValley region of the Northern California Bay Area. The collapse of the Twin Towers and the destruction at the Pentagon revealed the importance of sharing data about critical infrastructure when disaster strikes. TriValley emerged as an ideal setting for a coordinated regional GIS program.

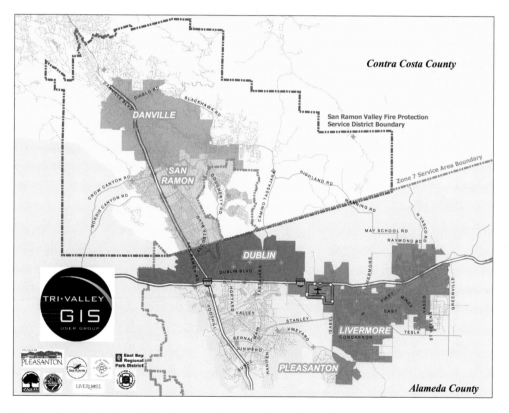

This map shows the Northern California communities that make up the TriValley area where a regional GIS network has improved communications and emergency preparedness.

Source: Rusty Wynn, City of Pleasanton, July 2006.

The TriValley area is vulnerable to earthquakes and wildfires, is a cargo gateway to the port of Oakland, and is home to a military base and two high-security nuclear weapons research facilities, the Lawrence Livermore National Laboratory and the Sandia National Laboratories. Any emergency event likely would involve multiple agencies, so a regional approach to prevention, mitigation, and response made sense.

Yet, communication and GIS cooperation among the multiple jurisdictions within the TriValley region remained sporadic at the start of the twenty-first century. The TriValley Business Council, which brings together business, government, and community groups for regional planning, was the catalyst for a GIS regional model. The council developed a vision for the TriValley GIS project in 2000, and the September 11 attacks heightened the urgency for implementation.

A workable plan

The business council's needs analysis showed that a regional GIS network would improve communication, emergency preparedness, economic development, impact analysis, and decision making while providing each participating agency with more information at less cost. Each public agency would use its existing GIS software while data from the multiple platforms would be readable for everyone involved. Starting in 2003, cities, special districts and other agencies signed the regional data sharing agreement. Once the legalities were settled, TriValley GIS members set their sights on merging disparate datasets, their most daunting technical challenge.

The TriValley GIS Project received technical assistance from Hewlett-Packard (HP) and ESRI to solve data-integration issues. HP provided a central homeland security data server that enabled various jurisdictions to share critical infrastructure data. This cooperative system complemented national and local homeland security and first-responder initiatives. The goal was to build an expandable architecture to support information sharing, analysis, and reporting within a secure, heterogeneous environment that disaster and emergency planners would use.

Study results in seamless map

A homeland security pilot study of the Livermore, Pleasanton, and Dublin areas resulted in a seamless digital map that included imagery, parcels, centerlines, and contours. The participating agencies, including the cities of Danville, San Ramon, the Zone 7 Water Agency, the San Ramon Valley Fire Protection District, and the East Bay Regional Parks District, also created regional land-use classifications and acquired regional high-resolution/high-accuracy orthoimagery.

The TriValley GIS Project met its interoperability goal, allowing any agency in the partnership to share datasets, some from disparate platforms. The project evolved toward Web-accessible views of the shared geospatial data.

Lessons learned from the TriValley model included the following:

- Appoint a dedicated project manager as a contact person for software, hardware, and data-integration issues.

- Involve decision makers from various groups such as public safety, city GIS, IT management, and the private sector.
- Representatives from multiple agencies must come together to break down information silos.
- Launch the GIS model in phases. Start with core data, then serve it to users to assure buy-in for the future. More data layers can be added later.

CASE STUDY

Kentucky's Chemical Stockpile Emergency Preparedness Program (KY CSEPP)

Few critical infrastructure sites pose a greater risk to the surrounding area than a military installation where tons of chemical weapons and thousands of explosive munitions are stored. The Blue Grass Army Depot in Kentucky is under a congressional mandate to eliminate its chemical stockpile, but the Commonwealth of Kentucky lacked a cohesive and effective way to visualize and manage all of the aspects of the emergency preparedness plan for the ten-county region surrounding the depot.

State officials realized that organizing vital information about evacuation routes and nearby hospitals could be expedited with GIS. But the predominantly rural region, which had long relied on telephone books and enlarged, laminated highway maps marked with grease pencils, had yet to fully enter the digital age and had little money to modernize. PlanGraphics, a Frankfort, Kentucky, geospatial consulting firm, teamed with Kentucky's Chemical Stockpile Emergency Preparedness Program (KY CSEPP) to obtain grant funds and implement an affordable, multiphase mapping support system.

Risk to the region

The Blue Grass Army Depot covers 14,600 acres of open fields and wooded hilltops and valleys in east central Kentucky where the nearest cities of Richmond and Lexington have a combined population of more than 300,000. The depot, built in the 1940s, stores about 55,000 explosive devices, including rockets and land mines, and about 523 tons of chemical weapon munitions containing mustard gas, VX nerve agent, and sarin, also known as GB. The Kentucky facility is one of eight chemical stockpiles in the United States, and Congress ordered the Army to eliminate the aging weapons at all eight sites under provisions of the international treaty on chemical weapons. Because the removal process could increase the risk of accidental leaks, the government further ordered stepped-up emergency preparedness to protect communities surrounding those sites.

With a $38,000 grant from the Federal Emergency Management Agency (FEMA) in late 2001, KY CSEPP officials launched an effort to computerize maps of the almost 3,000-square-mile area around the depot. In the first phase, PlanGraphics acquired computer and printing hardware, and ArcView software to build KY CSEPP's system. PlanGraphics

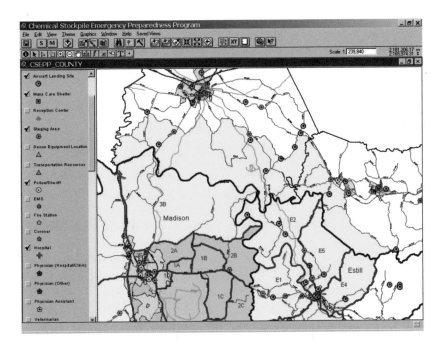

Kentucky CSEPP phase two thematic map showing Blue Grass Army Depot (green), response zones (pink and yellow), and the locations of nearly sixty different types of features and facilities.

Source: PlanGraphics and Kentucky CSEPP.

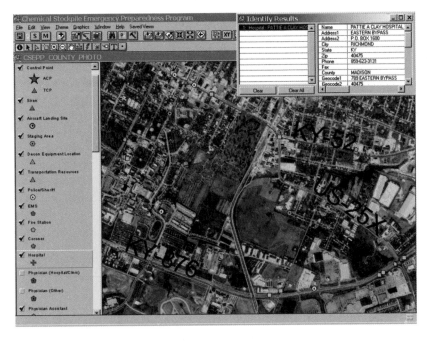

Phase two one-meter panchromatic imagery map of the City of Richmond in Madison County, Kentucky, showing various features and facilities with attribute detail on the hospital presented in the identify pop-up window.

Source: PlanGraphics and Kentucky CSEPP.

then sought geographic data from various state and local sources to develop basemaps for each county.

Getting started

The consulting firm assembled a consistent regional base using thematic data, one-meter aerial photos, ten-meter SPOT satellite imagery, and DRG topographic maps. The former Kentucky Office of Geographic Information provided much of the raster data needed to prepare imagery and topographic basemaps. Phase one concluded with a survey of each county's digital data on schools, hospitals, public safety assets, shelters, group facilities, and other important resources. But there was little or no data available, especially in the remote counties.

So, the second phase focused on filling the void by building a database from diverse sources. The Kentucky Board of Medical Licensure's records of all licensed physicians and physician assistants were used to partially populate one of many needed attribute databases. That source was also used to map the location of approximately two thousand medical professionals.

Similar datasets were found for pharmacies, veterinarians, daycare centers, nursing homes, and group-care facilities. Much of the information was on paper or in simple spreadsheet and text documents and had to be converted to a suitable digital format. Kentucky lacked a master address database for its 120 counties, so the KY CSEPP purchased addressing software to geocode locations with situs addresses. County disaster preparedness and public safety officials were particularly helpful in creating the database by identifying known locations on one-meter digital orthophotography for digitizing and providing GPS coordinates for such features as emergency landing zones.

After the data was collected and processed, PlanGraphics developed an ArcView project that organized the information in a format easily displayed and understood by non-GIS professionals so that users could quickly locate data by answering basic questions, performing simple queries, and clicking mapped features.

PlanGraphics developed a customized search capability to access the database by facility name, type, or location within a boundary feature, such as immediate response zones, cities, counties, or census tracts. The GIS was programmed with several buffering routines for locating features, such as evacuation sites, transportation resources, and highway control points. The ArcView project was further enhanced with the ability to locate and buffer an x,y coordinate position and identify and obtain more detailed information on features within that selected area.

Emergency personnel could map facilities affected by a chemical plume and access relevant data, such as available hospital beds or students at a school with contact names and phone numbers. Before GIS, most information was difficult to retrieve and use because it was simply written on paper notes, thumb tacked to bulletin boards, or filed in binders and cabinets.

Expanding functionality

In phase three, PlanGraphics expanded the ArcView project's functionality with additional search functions and by providing remote access from the Blue Grass Army Depot, county emergency operation centers, and federal government offices in other states. After purchasing and installing a dedicated server and ArcIMS software, PlanGraphics converted the ArcView project to an Internet mapping application to make maps and data available to various officials over a secure Internet site and to keep the GIS database synchronized. The system was designed to support response to all natural hazards beyond the scope of an accidental chemical discharge from the depot, including tornadoes, floods, and snowstorms.

GIS eliminated paper records and integrated multiple data files for planning, emergency response, and training, including the annual Blue Grass Army Depot exercise. As with many GIS projects, application utility can decline if the underlying data is not maintained, so PlanGraphics also prepared a database maintenance strategy and schedule to keep the system current.

In 2006, PlanGraphics completed the KY CSEPP project with data and Web-based applications serving first responders in the region, state emergency management personnel, and federal agencies, including FEMA and the U.S. Army. Using grant assistance for the Kentucky Science and Engineering Foundation, the mapping system was fully incorporated into a secure situation assessment and management portal to link with various other preparedness functions and applications, including traffic video surveillance, plume modeling, crisis incident management, evacuation routing, and incident redlining and secure messaging.

Among the lessons learned from the KY CSEPP project were the following:

- Mapping systems should be built sequentially and at a rate that supports adoption and ease of use by non-GIS professionals without mapping experience.
- Collaboration is key among numerous state, local, and not-for-profit agencies.
- Design development has to be affordable.
- Data maintenance is important.
- GIS and other emergency management applications must be integrated to provide all responders with common operational resources and perspective.

CASE STUDY

PureActiv video surveillance

Real-time situational awareness, the goal of any homeland security scenario, is the product of a breakthrough marriage of GIS with video surveillance. A system being introduced to airports, seaports, utilities, and other locations where critical infrastructure is priority one incorporates maps, aerial photos, street names, building locations and other GIS data with live video from strategically placed cameras.

This integration of ArcGIS with PureTech Systems' technology is called PureActiv, an automated outdoor surveillance solution that is able to focus on actual threats while minimizing false alarms. The system features automated camera positioning based on latitude and longitude coordinates to create detailed map displays that identify the locations of all security devices, cameras, and alarms, including the real-time graphics showing the camera's actual field of view. The GIS command and control user interface enables operators to zoom and pan maps to any level and control any device represented on the map. Clicking a camera icon immediately produces a live video feed from that camera.

Protecting water and waterways

The City of Phoenix, Arizona, where PureTech is based, chose PureActiv technology to protect its water treatment and storage facilities. The Minneapolis–St. Paul International Airports Commission and the Port of Seattle also installed PureActiv to meet their homeland security needs.

The PureActiv object detection and tracking module can be configured to detect objects of a user-specified size, speed, type, and direction while ignoring objects that do not meet those criteria. This module works with a variety of video cameras, including day, night, monochrome, color, and infrared.

Perimeter sensors may also be incorporated into the system, so when a disturbance is detected at a fence, cameras are immediately aimed at that location to catch the intruder in the act. GIS technology pinpoints location through latitude and longitude coordinates, enabling scene analysis and real-time situational awareness on which to base a security response.

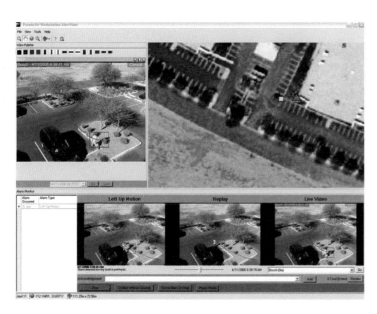

PureActiv Workstation Client showing the tracking of a threat on both the video view and the translation of that threat into the map view. Bottom window shows actual alarm pop-up window.

Source: PureTech Systems.

References

Directions Magazine. 2002. PlanGraphics, Inc., provides GIS support to Kentucky chemical stockpile. December 9. http://www.directionsmag.com/press.releases/index.php?duty=Show&id =6083&trv=1.

Dirksen, Helen, and Russ Johnson. 2003. *Project Management Plan for TriValley Homeland Security Information System Project.* Paper. October 7.

Fries, Jim. 2005. GlobalSecurity.org Web site. Description of Blue Grass Army Depot. http://www.globalsecurity.org/military/facility/blue-grass.htm.

Hewlett-Packard Development Company. 2005. *TriValley GIS Project.* June. http://h71028.www7.hp.com/ERC/downloads/4AA0-0787ENW.pdf.

Hilling, Bill. 2004. Disaster planning. *Government Technology* (June 3). www.govtech.net/magazine/story.php?id=90479.

Fire and Emergency Training Network. 2006. GIS: Providing the foundation for real-time intelligence. *American Heat* instructional series.

PureTech Systems Web site. www.puretechsystems.com.

RFDesign. 2006. Combining technologies to create a robust security system. May 16. http://rfdesign.com/news/PureTech-Puractive-solution/.

U.S. Department of Homeland Security. 2006. *National Infrastructure Protection Plan* http://www.dhs.gov/interweb/assetlibrary/NIPP_Plan.pdf.

U.S. Department of Homeland Security Science and Technology Directorate. The Executive Office of the President, Office of Science and Technology Policy. 2004. *The National Plan for Research and Development in Support of Critical Infrastructure Protection 2004.* http://www.dhs.gov/interweb/assetlibrary/ST_2004_NCIP_RD_PlanFINALApr05.pdf.

The White House. 2003. *Homeland Security Presidential Directive/HSPD-7.* http://www.whitehouse.gov/news/releases/2003/12/20031217-5.html.

3
Responding to complex emergencies

A levee breaks, a forest fire spreads, a bomb detonates, a chlorine cloud is unleashed. Such complex emergencies endanger lives and property, making a rapid response imperative. With GIS, the pressures of fighting the clock can be eased by reading a map. As evacuees in hurricane-ravaged New Orleans boarded makeshift rafts on flooded streets and filled the rafters of the Louisiana Superdome, technicians armed with GIS knowledge and equipment helped restore order.

Firefighters, emergency medical personnel, and law enforcement typically are the first responders to an incident. GIS and mobile computers provide critical information, such as floor plans, aerial imagery, and the presence of hazardous materials, that helps them size up the situation immediately. The fire service in California pioneered the incident command system (ICS) in the 1970s as an efficient way to integrate resources to quell a blaze. Before ICS, fire agencies found it extremely difficult to work together on major wildfires because of organizational and communication deficiencies.

The key features of ICS are the following:
- A unified approach to incident management
- A common terminology for clear communication
- A generic organizational structure that provides for interagency cooperation
- A chain of command with accountability
- Efficient use of resources

ICS is organized by subsections, allowing for staffing to be enlarged or reduced based on the needs of the incident. By the 1980s, all state and federal agencies had adopted the incident command system to manage wildfires, hazardous waste spills, and other emergencies.

Borrowing from the ICS model

The National Incident Management System (NIMS), the Department of Homeland Security's standardized approach to unify all levels of government for incident response, is based on the fire service model. By using standardized procedures, responders share a common focus. In any homeland security event, whether terrorism or natural disaster, all emergency teams follow a common protocol under the NIMS, which also sets criteria for personnel competence, training, exercises, and equipment. The hallmark of the incident command system is its flexibility to adapt to different conditions and circumstances.

Homeland Security Presidential Directive 5 (HSPD-5), issued on February 28, 2003, mandated that federal agencies adopt the NIMS. HSPD-5 also required state and local agencies to use the system as a condition of receiving federal funds. HSPD-5 stresses the need for technology by advocating interoperability standards, information management systems, GIS, and other supporting technologies. Since GIS has long been an operations and training component of wildfire management, it was quickly integrated into the NIMS for broader disaster responses.

Technical support for an incident management system is based at the incident command post. There, the incident command staff can call up digital maps that display the location, nature, and status of an incident. GIS also comes into play while tracking the progress of a fire, hazardous chemical release or other emergency or when assessing crucial resources, such as hospitals and shelters. GIS can model the incident to predict where it will spread and resultant consequences for the next hour or next several days. At complex multiday incidents, public safety personnel are given an incident action plan for each twelve-hour work period. The incident action plan consists of GIS-produced maps, including the following:

- The operations map showing the incident perimeter, how the incident is geographically divided into branches and divisions, the work assignment in each division, who is assigned, and so forth
- The hazard map showing areas where extreme hazards or unique conditions exist
- The logistics map showing roads and routes to access the incident and support facilities, such as fueling locations, general supply caches, medical support, and drop points

The emergency operations center supports the incident command post requirements and assures that government operations around the incident are functioning appropriately.

Key government department heads usually assemble at the emergency operations center to make unified decisions to support the incident and the community. This requires good information or real-time situational awareness. GIS is a critical component that provides a dynamic view of the incident, the surrounding community, and what actions are required to sustain operations. As the incident changes, or new incidents occur, GIS can dynamically update the map and provide detailed information about conditions.

Kinks in the system

Hurricane Katrina revealed the chasm between the standards of NIMS and the reality of the emergency response in New Orleans and throughout the Gulf Coast. As a White House report on lessons learned from the disaster acknowledged, "While we have constructed a system that effectively handles the demands of routine, limited natural and man-made disasters, our system clearly has structural flaws for addressing catastrophic incidents."[1]

The report called for a new "culture of preparedness" in which all government assets and resources are integrated and synchronized to ensure an effective national response to a crisis. To accomplish this requires stepping away from the bureaucratic view of a particular department or agency's institutional interests and instead building preparedness partnerships across federal, state, tribal, and local governments.

CASE STUDY

Hurricane Katrina

After Hurricane Katrina hit in August 2005, rescue workers toiled in the soggy trenches to save lives and recover bodies while other, less-heralded volunteers scrambled to make sense of the big picture under trying circumstances. These were the GIS experts who realized that consistent and timely mapping were crucial to the relief effort, whether it was positioning facilities and services, prioritizing infrastructure repair, or assessing losses. Much of the response came from out-of-state volunteers unfamiliar with New Orleans and the Gulf Coast region.

The GISCorps, established in 2003 as an arm of the Urban and Regional Information System Association (URISA), mobilized and coordinated hundreds of GIS volunteers along the Gulf Coast. Teams created storm-surge flooding models and produced street-level maps and search and rescue grids. Working around the clock, volunteers pooled aerial photos and other data sources to create thousands of feet of maps. They distributed various maps to enhance overall situational awareness for incident commanders and, with more customized maps, assist

1. The White House. The Federal Response to Hurricane Katrina: Lessons Learned. February 23, 2006. http://www.whitehouse.gov/reports/katrina-lessons-learned/index.html.

A touch of beauty amid the Hurricane Katrina devastation in the 9th Ward of New Orleans.

Photo courtesy of Marie Lynn Miranda.

emergency crews on the ground. Volunteers translated addresses into GPS coordinates for the U.S. Coast Guard helicopter rescues and maintained a Web site that mapped last-known locations of missing people.

Helping the helpers

Other geospatial applications included assigning equipment, organizing and deploying personnel, planning evacuations, modeling and tracking hurricanes, assessing damage, restoring infrastructure, and identifying emergency shelters. American Red Cross officials responsible for getting food, clothing, shelter, emotional support, and other essential services to displaced hurricane victims praised GIS technology for making their job easier. The ability to map and analyze priorities enabled the Red Cross to work quickly and efficiently while maintaining enough flexibility to meet changing needs.

In rural Mississippi, GISCorps volunteers brought GIS to a region that severely lacked technology and also guided the relief efforts of the Federal Emergency Management Agency (FEMA) and U.S. Army Corps of Engineers with innovative mapping techniques.

Spatial technologies at work along the Gulf Coast had to support a smorgasbord of communication systems and datasets. Mapping efforts transformed disparate datasets from many agencies into a form suitable for broad dissemination. GIS data fed computer models that provided emergency crews with maps, charts, and graphics they could follow, such as hazard zones subject to flooding, residential demographics, and clean-up calculations. Users could access the Web application to look at data for a specific area to find out the extent of damage to houses, property, and critical infrastructure. Such information formed a common operating picture that was vital to officials setting priorities for scarce resources.

GIS for the Gulf

A joint effort by the U.S. Geological Survey (USGS), the National Geospatial-Intelligence Agency, and the Department of Homeland Security led to GIS for the Gulf (GFG). Built on the GIS for the Nation data model, GFG merged data from various systems into a secure, Web-based repository that met immediate and long-term recovery needs and helped prepare for future hurricanes. Immediately after Hurricane Katrina hit, officials from various federal agencies knew that if the different GIS systems and datasets were to be combined, it required multiple-agency participation at all levels of government and private industry. Organizations were motivated to share, import, integrate, and synchronize

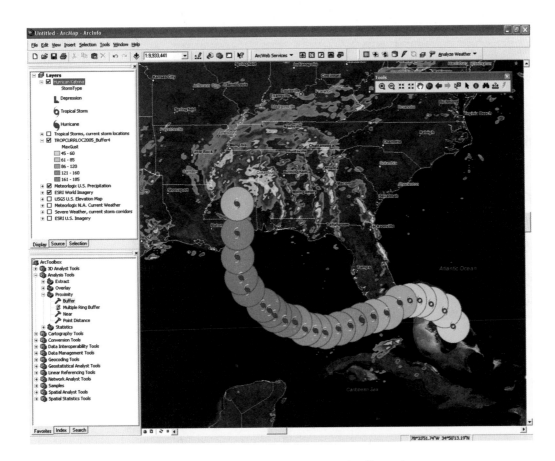

Hurricane Katrina weather data is tracked through ArcWeb Services and displayed in ArcMap.

Sources: USGS, U.S. Census; weather data courtesy of Meteorlogix.

After Hurricane Katrina hit,
New Orleans levees gave
way, flooding much of
the city.

Source: National Oceanic and
Atmospheric Administration.

The storm tossed around
boats at the marina in
New Orleans.

Source: National Oceanic and
Atmospheric Administration.

datasets into emergency operational databases. The comprehensive package was based on a standardized, multiscale model that provided a consistent view of data across jurisdictional boundaries.

The 2004 tsunami in Asia and other deadly disasters prompted U.S. government officials to focus on improving collaboration among various organizations. The ability to quickly leverage independent local datasets is particularly important in an emergency. In February 2005, the USGS explored an all-purpose data-sharing information model that would work for local, state, regional, and national needs. Six months later, Katrina struck and much of the groundwork had been set, so the collaborative model was applied to the affected area to assist emergency responders and recovery planners. The GFG database developed about sixty layers, including detailed parcel information and aerial imagery, combined with a suite of applications that allowed data to be viewed, analyzed, and manipulated through a password-protected Web portal.

The ESRI Web site contained a wealth of public information, such as hurricane advisories and response updates, maps, and images. A Hurricane Katrina Disaster Viewer showed data and imagery for the hurricane region and allowed users to create personalized maps and demographic reports. ESRI also offered Business Analyst Online to agencies involved with economic development and planning, trade associations, or any other party that wanted to study population demographics, housing, and businesses affected by Hurricane Katrina. Business Analyst Online combines GIS technology with

Hurricane Katrina Coastal Inundation: Gulf of Mexico

This map shows areas along the Gulf Coast inundated by Katrina flood waters.

Source: National Oceanic and Atmospheric Administration.

extensive and accurate demographic, consumer, and business data to deliver reports and maps over the Web.

FEMA's geospatial analysts mapping support helped first responders and emergency recovery personnel meet their mission. Highly accurate flood maps were created with lidar, the remote-sensing technology that uses lasers to measure distances to reflective surfaces, which also provided incident management personnel an accurate understanding of a very large disaster spread over several states.

CASE STUDY

Graniteville, South Carolina, train wreck

A predawn train wreck in rural South Carolina unleashed a cloud of deadly chlorine gas on January 6, 2005, mobilizing an emergency response that was planned and executed with the help of GIS. Nine people died and hundreds more were injured, but a well-coordinated evacuation plan spared thousands of others.

Graniteville was founded in the mid-1800s as a cotton-mill village known for its blue granite buildings. In 2005, the town was still home to several textile plants and a major tire

Aerial view of the Graniteville freight train disaster.

Source: Environmental Protection Agency.

Emergency workers in protective gear respond to the wreckage.

Source: Environmental Protection Agency.

manufacturing company. Those industries meant daily trainloads of chemicals and other materials passing through the region near the Georgia border.

At about 2:40 AM, a Norfolk Southern Railway freight train arrived in Graniteville and crashed into a parked locomotive because a rail switch had been improperly set. The collision ruptured one of the train's tankers carrying ninety tons of chlorine. Four other chemical tankers derailed but did not leak. The released chlorine vaporized, sending a huge toxic cloud over the wreckage. Graniteville, with a population of about 1,500, maintained a small volunteer fire department that was ill-equipped to respond to such a major emergency. So, the first-responders were dispatched from Aiken County, and they brought GIS with them.

Deadly chlorine gas

Five night shift workers at the Avondale Mills plant near the crash site and the train engineer died from inhaling the vaporized chlorine. Chlorine gas is extremely corrosive to metal parts and electronic components and deadly when inhaled by humans. It affects respiratory and central nervous systems and can damage the throat, nose, and eyes.

Dozens of local, state, and federal agencies converged on Graniteville. Aiken County's GIS provided critical spatial information to these emergency crews, including the status of the plume, the location of hazard zones, and the riskiest areas. They knew what protective equipment they needed and precautions they should take. The GIS also illustrated which areas required immediate evacuation and where roadblocks should go. All responders had a consistent, frequently updated, picture of the emergency.

Maps make the difference

Aiken County's GIS staff distributed hundreds of large maps to the various emergency crews to keep them updated during the two-week response. The GIS-generated maps displayed the crash location, the command post, cordoned-off areas, and danger zones. Those maps were invaluable to the emergency management personnel and assisted in saving lives by reducing exposure to the gas cloud. An estimated 4,000 to 5,400 residents were evacuated, starting with mandatory evacuations for those living within one mile of the crash site. The Aiken County Sheriff's Office evaluated the response and credited the availability of crucial GIS data for the speedy evacuation of those closest to the spill. The sheriff's report also praised GIS personnel for continuously distributing maps that supported emergency operations.

GIS also came into play when the University of South Carolina evaluated the emergency response, particularly evacuations. Spatial analysis using ArcGIS demonstrated that despite some confusion and disparities in how the evacuation proceeded, there was an exceptionally high evacuation rate and more than two-thirds of survey respondents had only positive observations about how the incident was handled.

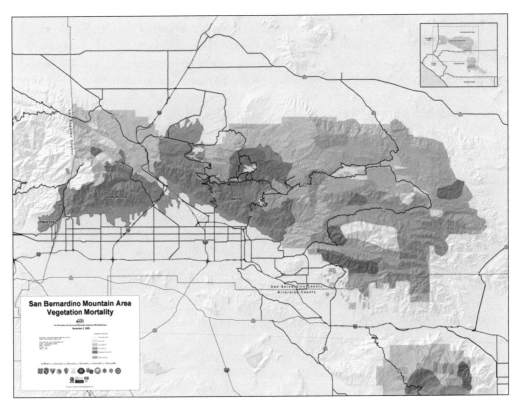

This map represents dying trees in the San Bernardino National Forest, an important factor of the fast spread of the fires.

Source: ESRI. Data courtesy of National Park Service, ArcUSA, TANA/GDT, U.S. Census, U.S. Bureau of Transportation.

CASE STUDY

California wildfires

An extended drought and hot, dry weather had turned Southern California's mountains into a massive tinderbox in 2003. A devastating firestorm seemed inevitable, and officials in Riverside and San Bernardino counties braced for the fight with spatial technology. When the fires did strike that fall, the planning paid off. Federal and local agencies had created MAST (Mountain Area Safety Taskforce) to coordinate a strategy based on frontline applications of GIS and GPS. MAST was formed by bringing together more than twenty-five agencies from all levels of government and the private sector to develop strategies for dealing with an enormous die-back of pine trees in the San Bernardino National Forest.

Two separate fires—the Grand Prix and Old, named for the roads at their origins—merged in late October in the San Bernardino Mountains, threatening homes and forcing evacuations. By the time the inferno was quelled in early November, 1,134 homes were destroyed but emergency preparations, especially MAST's evacuation plan drafted months earlier, saved thousands of others. GIS also assisted fire-suppression crews by tracking the

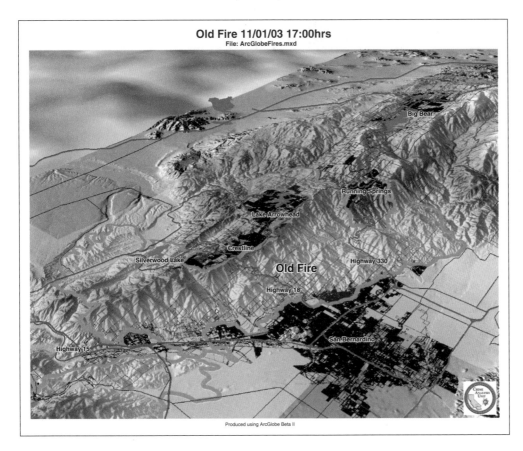

CalMAST and the San Bernardino County Sheriff used ArcGlobe to visualize a 3D flyover of the Old and Grand Prix Fire perimeters.

Source: ESRI. Data provided by CalMAST and USGS.

fire's progress with near-real-time imagery, identifying critical facilities, monitoring traffic, assessing damage, and much more.

Planning pays off

Drought, in combination with a bark-beetle infestation, killed nearly a million trees in the forests of the Riverside and San Bernardino counties, ramping up fire danger. MAST's first priority was to get the lay of the land. Relying on the GIS experts from its member agencies, MAST mapped the San Bernardino National Forest's topological data, hydrology, soil, vegetation, population, and road system. A MAST Web site became a key source of public information, including shelter locations and advice for protecting property, before and during the fire.

Handheld computers loaded with ArcPad GIS software helped firefighters process geographic data on fire movements practically in real time. The wireless technology enabled helicopter crews to punch in fire perimeter coordinates and transmit that data to ground

The 2003 Old Fire in the San Bernardino Mountains of California.

Photo by Christopher Doolittle.

personnel. Briefing and strategy sessions became clearer through 3D maps created with ArcGlobe that simulated the fire-ravaged environment. Fire commanders and crews could quickly see vegetation types, slope characteristics, and threatened resources by studying the ArcGlobe display.

This technology helped organize evacuations, locate fire lines, model fire progression, and predict fire behavior. This also allowed fire operations personnel to determine, based on slope, vegetation, roads, and proximity to critical protection values, where safe and effective fire suppression tactics could be deployed. Law enforcement also used GIS software to organize evacuations, prevent and investigate crime in the fire area, monitor traffic, and plan community reentry.

Maps and mobile technology

Large-scale maps were printed at information centers and evacuation points to brief firefighters, the media, and the public, and also to distribute tactical information to incoming fire crews. Mobile GIS technology aided fire crews as they assessed damage once the blaze was extinguished. They stored digital photographs and notes on their handheld computers and later transferred that data into MAST's main database. The detailed damage assessment expedited federal disaster relief funds and gave officials a leg up on the flooding and mudslides that were expected to follow with the winter rains.

In the years since the Grand Prix–Old disaster, real-time mobile mapping technologies that integrate GIS, GPS, and two-way radio have become faster, more accurate, and more useful for firefighters. With a wealth of information at hand—topographic maps, aerial photos, landownership boundaries, air and ground hazards, and weather conditions—fire crews are wiser and safer in the face of danger.

References

Aiken County Sheriff's Office. 2005. After-action report, Graniteville train wreck. January.

ArcNews. 2005. GIS supports hurricane response. Fall. http://www.esri.com/news/arcnews/fall05 articles/gis-supports.html.

———. 2005. GIS volunteers help in Mississippi's Hurricanes Katrina and Rita response. Fall. http://www.esri.com/news/arcnews/fall05articles/gis-volunteers.html.

———. 2005. USGS, NGA, and DHS collaborate to build "GIS for the Gulf." Winter 2005–06. http://www.esri.com/news/arcnews/winter0506articles/usgs-nga-dhs.html.

———. 2003. GIS helps response to California wildfires. Winter 2003–04. http://www.esri.com/ news/arcnews/winter0304articles/gis-helps.html.

Bishop, Frank. 2005. GIS in an emergency situation. Presentation at the twenty-fifth annual ESRI International User Conference, August 9–13, San Diego, California.

CACI Technologies Incorporated. 2004. Incident Management System Technologies white paper. http://www.maritimesecurityexpo.com/whitepapersarticles/Incident%20Management%20 Systems.pdf.

Carpenter, Katy, and Lucia Barbato. 2006. Katrina: GIS Volunteers to the Rescue. Presentation at the twenty-sixth annual ESRI International User Conference, August 7–11, San Diego, California.

CBS News online. 2005. S.C. train wreck gas cloud kills 8. January 7. http://www.cbsnews.com/ stories/2005/01/08/national/main665649.shtml.

Essex, David. 2006. DHS special report. FEMA maps out a better response. *Government Computer News* (June 19). http://www.gcn.com/print/25_16/41081-1.html.

Fire and Emergency Training Network. 2006. GIS: Providing the foundation for real-time intelligence. *American Heat* instructional series.

Government Technology. 2004. GIS solution helps Mountain Area Safety Taskforce quell Southern California blaze. December 29. http://www.govtech.net/magazine/channel_story.php? channel=19&id=92605.

Healthy GIS. 2006. American Red Cross uses GIS for Hurricanes Katrina and Rita efforts. Winter. www.esri.com/library/newsletters/healthygis/healthygis-winter2006.pdf.

Mitchell, Jerry T., et. al. 2005. Evacuation behavior in response to the Graniteville, South Carolina, chlorine spill. Quick Response Research Report 178, University of South Carolina.

National Incident Management System Web site. http://www.nimsonline.com/nims_3_04/index.htm.

Noland, David. 2005. Collision course. *Popular Mechanics* (April). http://www.popularmechanics.com/science/transportation/1513937.html.

Patterson, Tom. 2006. Using mobile mapping to manage wildfires. *Geospatial Solutions* (February). http://www.geospatial-online.com/geospatialsolutions/article/articleDetail.jsp?id=303344.

The White House. 2003. *Homeland Security Presidential Directive/HSPD-5.* http://www.whitehouse.gov/news/releases/2003/02/20030228-9.html.

GIS Profiles

This section recognizes four dedicated professionals who have put GIS to the test under difficult, even dangerous circumstances. GIS and homeland security, at its core, is about innovative people finding new ways to communicate vital information quickly and easily. Steven Robinson, Bill Hilling, Ron Langhelm, and Marie Lynn Miranda are shining examples of GIS leadership down in the trenches or, in Robinson's case, up in the air. It is through the efforts of these people and many others like them that GIS technology has become an indispensable, practical tool for promoting homeland security.

Steven Robinson
From tragic crash to technical triumph

Photo courtesy of Steven Robinson and ESRI.

Steven Robinson's career as a helicopter pilot for the Los Angeles City Fire Department ended after he survived a crash that killed four people in a rescue operation. But he has remained a dedicated firefighter who substituted technological ingenuity for flying skills to continue saving lives and property.

Robinson suffered head injuries in the 1998 crash, which was caused by a mechanical failure, and endured rigorous rehabilitation to briefly regain his pilot's license. A seizure in 2001 stemming from the head injuries ended his comeback attempt for good, but Robinson was determined to stay with the department's air operations team in some capacity. He took on a new role as an aerial firefighting and safety instructor and also camera and infrared operator, which kept him in the air.

Robinson soon realized that the UltiChart computer on board the fire department helicopters had potential far beyond telling pilots their position. The equipment could spatially reference ground features to calculate fire perimeters. While flying over fires, Robinson and the pilot would georeference the area, looking for hotspots, and radio the information to commanders on the ground. Robinson later added GIS to his assessment system, turning to ESRI for technical assistance layering geospatial data on high-resolution video images.

Working with Russ Johnson, ESRI's public safety manager, Robinson devised a system that blended dynamic fire-scene data with base data, such as census and street information, to help coordinate fire suppression as well as emergency planning and response.

The system could map the scope and speed of a fire, predict its potential path of destruction, and calculate the people, places, and things at risk. Incident commanders now had vivid displays with vital information to take the guesswork out of where, when, and how to deploy fire crews while maximizing their safety.

The forecasting feature of Robinson's system—a synthesis of data about weather conditions, terrain, and dry vegetation fueling the fire—proved invaluable in September 2005 during a 24,000-plus acre fire northwest of Los Angeles. Robinson calculated the fire's path and burn rate based on fire perimeter measurements and other data. That information served as an early warning system that helped fire officials stop the blaze from jumping a major highway, U.S. 101, toward Malibu and the Pacific Ocean. The fire resulted in no fatalities.

While responding to a hazardous materials incident, Robinson used a high-definition camera from the helicopter to read the dangerous contents of a fifty-five-gallon drum, providing firefighters with critical safety information before they approached the scene. Robinson has continued to seek ways to improve the system, including expanded use of video and Internet technology.

It took a tragic accident that forced a helicopter pilot to the ground for an innovative firefighting and disaster-assessment system to get off the ground and spread like wildfire.

Bill Hilling
Leading rural Kentucky to the digital age

Photo courtesy of Bill Hilling.

When Bill Hilling became planning supervisor of the Chemical Stockpile Emergency Preparedness Program (CSEPP), he was determined to achieve his predecessor's dream of replacing old enlarged paper highway maps with a computerized map system that would help safeguard Kentucky's citizens should there be an incident at the Blue Grass Army Depot.

As a retired Army officer and Vietnam veteran whose assignments included artillery and armor, Hilling knew maps to be an indispensable part of the military. The largely rural character of the region surrounding the depot and its limited use of technology presented challenges. Hilling knew it would take time to develop a regional mapping initiative and train officials in ten counties to use the new system. Since federal dollars would help pay for the project, Hilling broke it down into phases that could be more easily funded, managed, and locally adopted. New information would be gathered at each phase to add layers to the basemap and provide more functions within the application, and each phase would build on the previous one that counties had started to use. Each subsequent phase would give end users quicker access to greater amounts of useful information.

Flexibility in organizing and presenting information and introducing the technology was another key to the project's success. With its flexible stages, the project continuously accepted new datasets and map layers and easily migrated from ArcView to ArcIMS as Web-based mapping grew more popular and the CSEPP counties established more reliable communication networks. By moving to ArcIMS, all jurisdictions in the CSEPP footprint could access the GIS mapping program remotely, and consistent maps and underlying information could be readily displayed in each jurisdiction's emergency operations center (EOC).

Hilling met his goal to make mapping available to all CSEPP users—most of whom had no GIS familiarity when the project first began—in less than three years. Many people would have been content to stop there, but Bill Hilling's background as an Army officer and his awareness of the coordination problems following the September 2001 attacks motivated him to move further ahead. The fourth project phase created a common operational picture by integrating mapping, plume modeling, alert notification, incident management, evacuation routing, and other emergency management functions and resources in a single portal. With grant support from the Kentucky Science and Engineering Foundation and in-kind matching assistance from both CSEPP and PlanGraphics, which was CSEPP's contractor throughout all phases of the project, a fully integrated secure situational awareness was developed and first deployed in late 2005.

For the very first time, the CSEPP counties now had the capability to easily, quickly, and securely access thematic maps, aerial photos, and topographical maps of their jurisdictions combined with detailed information on more than sixty-two data layers covering every resource that could be involved in a CSEPP incident. More important still, the GIS data could be readily combined with and contribute to other independent software applications and decision-making resources to give emergency managers and first responders a comprehensive view of the CSEPP region during training exercises and a real incident.

After a shaky start and many hurdles overcome with flexibility, patience, and a clear vision of the end goal, the CSEPP mapping program went online, providing vital information for the safety of numerous communities and tens of thousands of citizens in the CSEPP region. Hilling credited his predecessor for the original idea and PlanGraphics for building the datasets and applications. But it took Hilling's leadership, persistence, and understanding of the region and the need for the mapping system to move an intriguing concept to a practical working reality.

T. James Fries of PlanGraphics was principal contributor to this profile.

Ron Langhelm
FEMA's GIS point man at disaster scenes

Photo courtesy of Ronald Langhelm.

Armed with an incredible mission and a geospatial vision, Ron Langhelm has been on the front lines of some of the biggest disasters in recent U.S. history, including the attack on the World Trade Center in 2001 and Hurricane Katrina in 2005. Langhelm has been supporting FEMA's Emergency Response Teams on events across the country for over ten years. In this capacity, he has served as the geospatial lead for federal response efforts, deploying spatial technology at disaster scenes. In addition to his FEMA role, Langhelm is now an associate with Booz Allen Hamilton, the Virginia-based strategy and technology consulting firm.

Lessons learned from crisis situations managed by Langhelm have been carried from event to event, advancing the homeland security applications of GIS technology while also underscoring the need for human interaction and cooperation. In the early response phase of any event, human interaction can pave the way to incredible successes. Langhelm has seen that, as humans, we rely on our gut instincts in a crisis setting. This critical time frame is where partnerships are built out of common needs and a simple trust in one another. These disaster-based relationships can maximize the potential of limited geospatial resources, remove data-sharing politics, carry us toward a new potential, locate needed supplies, and just keep us watching out for each other, according to Langhelm.

Langhelm played a key role in the aftermath of the September 11 attack on New York City, coordinating FEMA's support to state and local emergency GIS efforts. Earlier that year, Langhelm led geospatial support to the recovery operations after the Nisqually Earthquake in Washington State. Later, he was involved in GIS operations to secure the 2002 Winter Olympics in Salt Lake City and led the GIS effort recovering debris from the disintegration of the space shuttle Columbia over Texas in 2003. After Hurricane Katrina struck in 2005, Langhelm was deployed to Louisiana where he directed FEMA's GIS operations in the massive response and recovery.

Each disaster Langhelm responded to has presented unique challenges, but all events have required quick decision making based on past experiences, a dedicated staff, and a wide range of geospatial data and technology. The combination of GIS and remote sensing assisted emergency crews in Louisiana in rescuing trapped victims, recovering remains, evacuating citizens, recovering historical items, assessing impact, and briefing many government leaders. Similarly, at Ground Zero in New York City, GIS experts helped decision makers monitor search teams, medical units, and mobile offices while ensuring the safety of personnel from fires, hazards, and debris. GIS staff used aerial imagery, lidar data, forward-looking infrared radar (FLIR) data, and GPS-derived datasets to support the aggressive search efforts as well as the recovery.

Marie Lynn Miranda
A caring response to overwhelming destruction

Photo courtesy of Les Todd of Duke University Publishing.

Marie Lynn Miranda's knowledge of GIS and compassion for people, especially children, drew her to the Gulf Coast with other researchers to study the health impacts of Hurricane Katrina. The Duke University environmental science professor was instrumental in building a Web site loaded with GIS data that helped emergency workers and public health investigators alike.

Miranda was keenly aware of the long-term health risks in New Orleans and elsewhere along the Gulf Coast, having researched the spike in North Carolina's asthma rates after Hurricane Floyd caused massive flooding in 1999. She found that residual moisture even after cleanup provided an ideal environment for mold spores, which causes asthma and other respiratory problems in children. As director of the Children's Environmental

Health Initiative at Duke University, Miranda used GIS analysis in researching the risks posed by children's exposure to lead, allergens, pesticides, and industrial contaminants.

After Hurricane Katrina hit in 2005, Miranda joined researchers from the University of California at San Diego, San Diego State University, the University of Kentucky, Columbia University, the Research Triangle Institute, and other academic centers to create maps for the National Institute of Environmental Health Sciences' Web site. She led the effort to integrate a wealth of existing, nonconfidential GIS data to aid public health and environmental workers in the field

The Web site helped assess environmental hazards caused by Hurricanes Katrina and Rita and expedited cleanup efforts. Online satellite maps plotted the locations of potential toxic sources, including chemical plants and refineries, in proximity to flooded areas. Interactive features showed people their potential exposure to toxic chemicals and radioactive material from medical facilities. Up-to-date information on safety and hazardous waste cleanup training for thousands of workers and a list of environmental health resources for medical responders also was available on the Web site.

GIS was expected to play a major role in studying the long-term health consequences facing people exposed to the contaminated floodwaters in the Gulf Coast.

"While we all sent in checks to emergency relief organizations, helping to construct the hurricane response Web site and working with first responders felt like good, hard, constructive work that we could do in the face of almost overwhelming destruction," Miranda said.[1]

References

Basgall, Monte. 2005. Analysis of Katrina's health, environmental effects to be aided by website with layers of data. Duke University press release. September 8. http://www.dukenews.duke.edu/2005/09/GIS.html.

Carroll, Chris. 2004. Trail of tragedy. National Geographic (February). http://magma.nationalgeographic.com/ngm/0402/resources_geo.html.

Healthy GIS. 2006. American Red Cross uses GIS for Hurricanes Katrina and Rita efforts. Mapping, GIS analysis, and Web services help American Red Cross outreach. Winter. http://www.esri.com/library/newsletters/healthygis/healthygis-winter2006.pdf.

Kinzie, Susan. 2005. The best, brightest offer aid. Colleges share knowledge, skills. *The Washington Post,* Sept. 19. http://www.washingtonpost.com/wp-dyn/content/article/2005/09/18/AR2005091801169.html.

Langhelm, Ron. 2002. The role of GIS in response to WTC: The first 30 days. Paper presented at the ESRI International User Conference, San Diego, Calif., July 8–12.

McKay, Jim. 2004. Flying high. *Government Technology* (December 2). http://www.govtech.net/magazine/story.php?id=92317&issue=12:2004.

1. Healthy GIS. 2006. Research group prepares GIS data for Katrina/Rita response portal. Winter. www.esri.com/library/newsletters/ healthygis/healthygis-winter2006.pdf.

Morris, Tony. 2007. Fire watcher. LAFD's GIS mapping system. *Vertical Magazine* (February/March). http://www.verticalmagonline.com/.

National Institute of Environmental Health Web site. HIEHS hurricane response. http://www-apps.niehs.nih.gov/katrina/.

Newcombe, Tod. 2006. A new angle. *Emergency Management Magazine* (May). http://www.emergencymgmt.com/story.php?id=99434&story_pg=2.

Valencia-Martinez, Angie. 2006. Mapping system predicts fire path. Pilot develops new technology. *Los Angeles Daily News,* July 24.

4

Preparing for disease outbreaks and bioterrorism

Nothing seems scarier than the unseen enemy, as borne out by the anthrax panic of 2001 and the continuing threat of virus pandemics. Monitoring disease, combating bioterrorism, and protecting the food supply all involve locations that can be mapped and analyzed with GIS.

Public health officials look under the microscope to understand the nature of disease while GIS provides them with a wide-angle lens for preventing the spread of disease. Using geospatial technology, health officials can monitor problems and formulate a response to avoid or mitigate catastrophe. GIS can map spikes in emergency-room visits by people sharing similar symptoms and provide an early warning about a public-health problem or bioterrorist attack.

Disease surveillance is mainly about geography, figuring out where people are exposed and predicting which areas are at risk of exposure depending on climate, insect infestation, and other factors. Epidemiologists with access to GIS technology can pinpoint the origin of disease, track its spread, and minimize the damage. The U.S. Centers for Disease Control and Prevention, the World Health Organization, the Pan American Health Organization, and national health ministries all endorse GIS as a vital public health tool.

Integrating spatial data

Homeland Security Presidential Directive 10 (HSPD-10), issued on April 28, 2004, sets the framework for the nation's biodefense. As bioterrorism remains a constant threat, public health officials strive to achieve a real-time, fully integrated GIS to track disease. Integration is the key because of the numerous points of data collection and multiple layers of care providers within the healthcare field.

Providers on the front lines of the battle are responsible for recognizing symptoms indicative of a bioterrorist attack or disease outbreak. Health and homeland security officials realize that there is little time to confirm an act of bioterrorism and isolate the organism before the disease spreads to more victims.

After the events of September 11, Congress passed the Public Health Security and Bioterrorism Preparedness and Response Act of 2002 (the Bioterrorism Act), which President George W. Bush signed into law on June 12, 2002. The act added controls on biological agents, added safety and security measures affecting food, drug, and water supplies, and promoted the development of countermeasures to bioterrorism. It also expanded grant opportunities for state and local government. Some jurisdictions used the grant money to train health professionals in geospatial mapping and analysis and to incorporate GIS and GPS technology into healthcare systems.

Disease outbreaks tracked

Several states adopted GIS as the centerpiece of their surveillance programs after mosquito-borne West Nile Virus reached North America in 1999. When the Severe Acute Respiratory Syndrome (SARS) virus hit Hong Kong in 2003, a Web-based GIS application enabled the public to track the spread of the disease and its eventual eradication. This easily accessible information was credited with helping to minimize public panic over SARS.

GIS also facilitates health preparedness by analyzing demand for hospital beds, medical staff, and specialized services in response to an incident and organizing the logistics of moving patients and deploying medical resources in a community. GIS can model quarantine and evacuation scenarios in the face of a spreading threat.

Preparedness lessons learned from Hurricane Katrina in 2005 were applied to the medical field, where information systems needed upgrading to provide timely, accurate information to decision makers facing emergencies. The value of GIS in an early warning system could be expanded in conjunction with advances in radiation detectors and similar sampling devices. Such a system would measure contamination at various test locations and extrapolate the data to determine radiation concentrations in a broader area.

Food supply at risk

Health preparedness in a homeland security context also means protecting the nation's crops and livestock from disease and bioterrorism. Homeland Security Presidential Directive 9 (HSPD-9) issued on January 30, 2004, established a national policy to defend agriculture against terrorist attacks, major disasters, and other emergencies. With its tracking and traceability functions, GIS is a critical component of a platform to safeguard the nation's highly vulnerable food supply. Concerns about mad cow disease and similar threats prompted a nationwide initiative to track livestock using GIS and other geospatial technologies.

The National Animal Identification System (NAIS) eventually will identify all livestock and premises that have had contact with an animal disease within forty-eight hours after discovery. The NAIS is designed to limit the scope of such outbreaks and ensure that they are contained and eradicated as quickly as possible. By 2009, federal agriculture officials expect NAIS to encompass a national database of geographically identified sites where animals are born, where they pass through, and where they are slaughtered. The identity of each animal and animal herd also would be entered into a national database. GIS would organize and visualize the data to forecast the spread of pathogens in livestock and help contain outbreaks.

Geography has long been an important factor in understanding and predicting the spread of disease. With biological warfare a legitimate threat, GIS is helping to guide the public health system's transition from response-driven toward a new paradigm of surveillance, prevention, and response.

CASE STUDY

Kansas cattle tracking

Homeland security isn't just about halting terrorist attacks or bracing shorelines against the ravages of nature. Protecting the nation's food supply also is a crucial mission. Mad cow disease (bovine spongiform encephalopath), bovine tuberculosis, and other public health threats are tied to the mobility of livestock. The National Animal Identification System (NAIS), a joint voluntary effort by the beef industry and animal health officials, has set a goal of identifying all animals and premises that have had contact with a disease within forty-eight hours of discovery.

A national program seeks to identify and track cattle and other livestock to prevent the spread of disease.

PhotoDisc/Getty Images.

Quickly tracing infected and exposed animals can reduce the impact of disease outbreaks by ensuring they are contained and eradicated. Tracking also minimizes the number of animals that must be destroyed. The U.S. Department of Agriculture (USDA) hopes to have the NAIS fully functioning by 2009. Some skeptical ranchers have raised concerns about the security of the data and investing in technology that soon could become obsolete, but others have enthusiastically participated and are anxious for results.

Kansas tests tagging

To test the feasibility of monitoring cattle movements, the Kansas Animal Health Department (KAHD) launched a pilot program in July 2005 that combined the technologies of GIS, GPS, Internet-based mobile communications, and radio frequency identification (RFID). Kansas State University cattle researchers teamed with state animal health officials to develop the program with an $809,000 USDA grant. Kansas, as the nation's hub for millions of cattle shipments originating in other states, was ideally positioned for the project.

The researchers applied RFID ear tags to cattle and equipped trucks with the technology to read the tags and transmit the data. The trucks transported the cattle to meatpacking plants. At each transfer site where cattle was loaded and unloaded, each animal's RFID data was combined with the GPS-based date/time/location stamps and transmitted through a wireless connection in near-real time, to an information management system. This process, when fully operational, would allow for an animal to be tracked from the farm where it was born, through the marketing processes, and finally to the slaughterhouse.

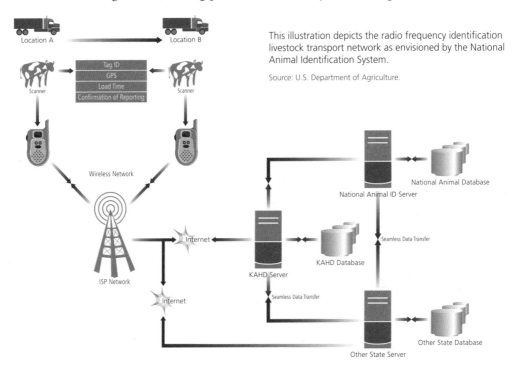

This illustration depicts the radio frequency identification livestock transport network as envisioned by the National Animal Identification System.

Source: U.S. Department of Agriculture.

Sites registered

The nationwide identification program requires ranches, feedlots, and other livestock sites to be registered to facilitate recordings of animal movements, such as ownership changes, transfers to areas where herds are commingled, interstate shipments, and slaughter.

Many states and tribal jurisdictions are developing registration systems. To qualify for registration, a livestock site must provide to the NAIS location information in the form of latitude/longitude, section/range/township, or 911 address. In Kansas, that data is then loaded into the Kansas Department of Emergency Management GIS, which provides geocoding capabilities for the address information. The system, which uses ArcSDE and ArcIMS software with an Oracle database, allows for spatial queries to locate all sites within a given area or map quarantine areas based on the spread of infection. State officials expect it to take years to register some 66,000 sites.

The Kansas project was the first to test a truck-based animal tracking system. The USDA initially allocated $11.6 million to test tracking ideas in other states, including proposals to improve cattle branding methods.

The seven-month pilot project in Kansas involved eight drivers who hauled 4,516 head of cattle and 175 head of swine. The results were hindered by equipment glitches and human error as only about 50 percent of the animal identification numbers were captured in the field. At best, under controlled laboratory conditions, the success rate topped out at 75 percent.

Nevertheless, in a report on the project's findings, researchers expressed confidence that the system would reach its full potential as RFID and related technology advances. Among the recommendations cited in the report were improved on-board equipment and better training for drivers and commercial companies providing data.

The report also underscored the key role the transportation industry will play in the success of the NAIS. Incentives, whether through tax breaks or subsidized equipment and education, would appear to be essential for livestock haulers to buy in to the animal tracking program, the report concluded. Meanwhile, voluntary identification has been gaining some momentum among suppliers who command premium prices because they can trace livestock movements from birth to slaughter.

CASE STUDY

Pennsylvania's West Nile virus surveillance

West Nile virus got its name from an area of Uganda where the disease was first detected in 1937. The virus, transmitted by mosquitoes, did not reach North America until 1999 but then quickly spread throughout the United States, initially killing seven people in New York City. Slowing the advance of West Nile required the speed and sophistication of GIS, a solution Pennsylvania officials adopted in 2000. Pennsylvania's health and environmental protection departments, working with other state and local agencies, developed a network covering all sixty-seven counties that involved trapping mosquitoes, collecting dead birds, and monitoring animals and people.

The virus can cause deadly encephalitis, an inflammation of the brain, or meninoencephalitis, an inflammation of the spinal cord and brain. But most people exposed to the virus show no ill effects and only some have mild flu-like symptoms. The elderly and people with compromised immune systems run the greatest risk of developing severe illness. Corvids, such as crows and jays, are most susceptible to the virus, so dead birds from that family are an early indicator of West Nile's presence. Horses are the only animals that can be vaccinated for West Nile virus and also are most susceptible to the illness. Like people, horses contract the virus through infected mosquito bites.

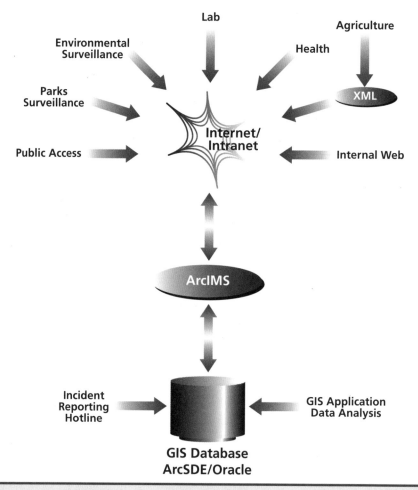

GIS is central to a West Nile virus surveillance program.

Source: ESRI.

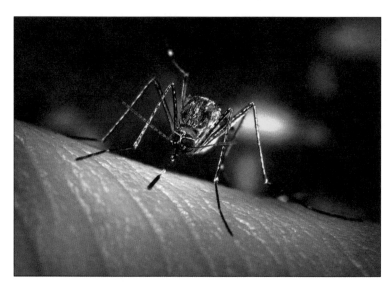

The *Culex tarsalis* mosquito is among the main vectors of West Nile virus in the United States.

Source: James Gathany, Centers for Disease Control and Prevention.

Closing the communication gaps

Learning from the 1999 virus outbreak in New York, Pennsylvania worked to close communication gaps by establishing the first multiagency West Nile virus reporting system based on GIS. The system tracks virus information spatially and stores data in a GIS central repository. The GIS marks the location of blood samples taken from people, horses, birds, and sentinel chickens. (Chickens produce antibodies in their blood when bitten by disease-carrying mosquitoes, so they serve a warning function.)

Mapping the location of mosquito breeding areas is critical for any subsequent virus control measures. The process is accelerated in the field with ArcPad software loaded on handheld computers. Workers immediately map locations and enter other sample information into the computer instead of using handwritten notes that must later be interpreted and then logged back at the office. A Web application facilitates access to data from state laboratories.

Current information posted

The tracking system features a bar code identification number placed on sample bottles. This number is used when field staff enters data into the handheld computer. When field data is uploaded, a quality-assurance/quality-control program verifies data accuracy; only verified data can enter the central database. In the laboratory, mosquito samples are identified, counted, and categorized by species and the information is added to the database. An internal Web server automatically checks the external database and retrieves any new data. This secure site is updated at noon daily and has hundreds of government users who review and discuss activity and decide where spraying or other control measures are needed. Public information, such as summaries prepared for each county, is available on the program's Web site by 2 PM daily. The site, www.westnile.state.pa.us, shows a map of Pennsylvania and data tables with county outlines so users know where West Nile virus has been reported.

Pennsylvania's GIS surveillance system could be applied to monitor other diseases and processes. For example, data uploaded from emergency medical services, physicians' offices, local health departments, public safety departments, and environmental monitoring stations could be incorporated into the system to protect against bioterrorism and respond to disasters.

CASE STUDY

New York City's syndromic disease surveillance

When it comes to detecting and stopping the spread of disease and bioterrorist attacks, GIS technology takes center stage. Public health and homeland security officials alike value surveillance systems that monitor, integrate, and analyze data on diseases and terrorism agents, whether biological, radiological, or chemical. During an attack or outbreak, the source must be quickly identified for an effective response. GIS, with its ability to transform real-time data into an easily understood spatial format, is a critical component of a disease surveillance system.

Following the September 11 attacks, the New York City Department of Health and Mental Hygiene (DOHMH) stepped up its bioterrorism alert by implementing an emergency department syndromic surveillance system with assistance from the Centers for Disease Control and Prevention. The system, based on data gathered daily from emergency departments, was designed to recognize an increase or clustering of syndromes that might reflect a natural or intentional disease outbreak.

New York City's emergency department syndromic surveillance system relies on data gathered daily from emergency departments.

S. Meltzer/PhotoLink/PhotoDisc/Getty Images.

This intensive monitoring, which included on-site staffing for data collection and entry, continued for weeks following September 11 and detected no outbreaks. An automated syndromic system subsequently was developed to monitor emergency department triage log data. Daily analysis that used to take an hour each day required no user intervention once the system was automated. About 90 percent of New York City's emergency department visits were covered under the system.

Symptoms key to detection

Traditional disease surveillance efforts were based on patient visits to doctors' offices and laboratory test numbers, and it usually took days or weeks to detect an aberrant pattern signaling an outbreak. But with a syndromic surveillance system, time is of the essence and symptoms and complaints, not the diagnoses, matter most. Early signs of many infectious diseases can be very similar and nonspecific, such as a cough or diarrhea. Delaying a response for a definitive diagnosis could cost lives. Respiratory and fever syndromes are monitored for early warning signs of bioterrorism while diarrhea and vomiting syndromes can indicate gastrointestinal outbreaks possibly due to contaminated food or water.

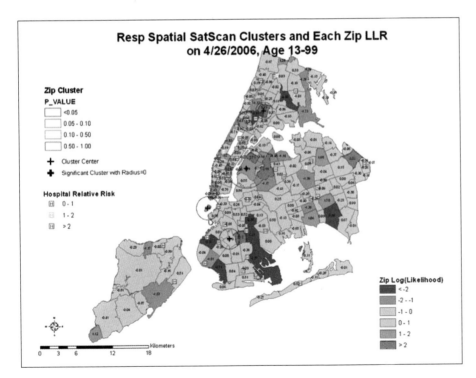

The city's health department uses spatial analysis software to map respiratory syndromes by ZIP Code and hospitals.

Source: Chris Goranson, New York City Department of Health and Mental Hygiene.

Besides hospital emergency department data, other key sources for the syndromic surveillance system are 911 calls and pharmacy sales. DOHMH started monitoring 911 calls in 1999, looking for signs of events ranging from food-borne illnesses to bioterrorist attacks. Over-the-counter medication sales would indicate people sick enough to buy a remedy but who may not require emergency treatment. City health officials relied on the surveillance system in 2003 to predict a flu outbreak two weeks before it hit, giving the public and medical providers time to prepare.

Tailoring technology

The backbone of the department's automated system is SAS SaTScan analytical technology integrated with ArcGIS geospatial software to streamline syndromic data analysis, create better maps, and cut response times to outbreaks. DOHMH officials developed a way to complement the strengths of SaTScan's spatial cluster analysis with ESRI toolkits and extensions using Python open-source scripting language.

As a result, syndrome mapping has been fully automated so that a disease analysis that used to take hours was reduced to minutes with a corresponding improvement in response to outbreaks. The process also enabled users unfamiliar with GIS tools to run the analysis and produce standardized maps.

DOHMH has emphasized the value of GIS throughout the department, forming a GIS working group to guide the transition from a shared network drive to an enterprise geodatabase with ongoing recommendations on GIS strategies. The department's GIS Center securely stores spatial data that connects to GIS support systems. The DOHMH's Emergency Awareness Application finds and tracks environmental hazard areas for emergency response. The department also equips users in the field with handheld devices to gather real-time spatial data.

Syndromic surveillance systems are the first line of defense against catastrophic outbreaks and an essential component of bioterrorism preparation. With a centralized data repository tied to existing data stores throughout DOHMH and with the GIS working group providing oversight and direction, GIS promises to play an increasingly important role in a multitude of public-health-related situations.

References

ArcNews. 2003 New York City uses GIS for surveillance of bioterrorism and disease. Fall. http://www.esri.com/news/arcnews/fall03articles/new-york-city.html.

————. 2000. Pennsylvania combats West Nile Virus with ArcPad, Internet. Winter 2000–01. http://www.esri.com/library/fliers/pdfs/penncombats.pdf.

Brown, Justine 2004. Prevention is the best medicine. *Government Technology* (March 2). http://www.govtech.net/gt/89565.

Coleman, Kevin. 2003. GIS, information technology, and biotech take center stage in supporting homeland security. Location Intelligence (April 11). http://www.locationintelligence.net/articles/345.html.

———. 2004. Bioterrorism and the food supply. *Directions* (October 1). http://www.directionsmag.com/article.php?article_id=667.

Conrad, Eric R. 2001. Tracking diseases with GIS. *ArcUser* (July–September). http://www.esri.com/news/arcuser/0701/wnvirus.html.

Davenhall, William. 2003. Preparing for large-scale events. *ArcUser Online* (October–December). http://www.esri.com/news/arcuser/1003/largescale1of2.html.

GeoIntellligence. 2005. Kansas tracks livestock in transit. September 1.

Goranson, Christopher. 2005. Disease surveillance and response systems. Presentation at the twenty-fifth annual ESRI International User Conference, August 9–13, San Diego, California.

Heffernan R, F. Mostashari, D. Das, A. Karpati, M. Kulldorff, D. Weiss. Syndromic surveillance in public health practice, New York City. Emerg Infect Dis [serial on the Internet]. 2004 May [date cited]. Available from http://www.cdc.gov/ncidod/EID/vol10no5/03-0646.htm.

Hegeman, Roxana. 2005. Animal ID program combines GPS, cell tech, RFID. Associated Press/ *USA Today* (February 1). http://www.usatoday.com/tech/news/surveillance/2005-02-01-cowtracks_x.htm.

K-State Research and Extension. 2004. Kansas animal health, K-State researchers awarded $805,000 to develop animal ID system. Press release. August 19. http://www.oznet.ksu.edu/news/sty/2004/animal_id081904.htm.

Lu, Jingsong, Christopher Goranson, Kevin Konty, and Farzad Mostashari. 2006. Automated creation of high-quality maps using SAS and Python. Presented at the 2006 Syndromic Surveillance Conference. October 18–20.

Pennsylvania Department of Health. 2004. Pennsylvania launches West Nile Virus surveillance program. Press release. April 28. http://www.dsf.health.state.pa.us/health/cwp/view.asp?A=190&Q=237401&pp=12&n=1.

Pennsylvania's West Nile Virus Surveillance Program Web site. http://www.westnile.state.pa.us/.

Peters, D. System Design Strategies, An ESRI Technical Reference Document. March 2006. Available at http://www.esri.com/library/whitepapers/pdfs/sysdesig.pdf.

RFID Journal. 2003. Can RFID protect the beef supply? January 5. http://www.rfidjournal.com/article/articleview/722/1/1/.

Stewart, Mary Ann. 2005. Tracking cattle in the heartland. *Geospatial Solution* (September 1). http://www.geospatial-online.com/geospatialsolutions/article/articleDetail.jsp?id=177059.

Teagarden, George. 2006. Final report. Kansas Animal Health Department USDA NAIS Pilot Project. Animal Transport RFID Validation. Kansas Animal Health Department. June.

The White House. 2004. *Homeland Security Presidential Directive/HSPD-9*. http://www.whitehouse .gov/news/releases/2004/02/20040203-2.html.

———. 2004. *Homeland Security Presidential Directive/HSPD-10*. http://www.whitehouse .gov/news/releases/2004/04/20040428-6.html.

5
Securing complex events

Be it a sporting event, political convention, or entertainment spectacle, large-scale gatherings attract widespread attention and, as potential terrorist targets, test the capabilities of geospatial technology. Complex events succeed when GIS processes don't go beyond preparedness and prevention, which was the case at the 2002 Winter Olympics in Salt Lake City and at both the Republican and Democratic national conventions in 2004. GIS helped security officials identify and monitor vulnerabilities and plan for emergency response in case something went awry. Nothing did.

As with natural disasters, complex events require GIS experts to work as a team, sharing data, formulating a common operating picture, and processing as much information as possible in real time, usually with secure, Web-based applications. Because national and international events involve a myriad of agencies unaccustomed to working with one another, planning and coordination is a must. In Utah, where tens of thousands of visitors attended Winter Olympic events, the dynamic data features of GIS technology monitored weather, radiation, and disease, as well as the comings and goings of athletes, contributing to precise situational awareness.

Security for large-scale events was redefined after September 11, adding new players to the mix such as the Federal Emergency Management Agency, and federal customs and immigration agencies. The National Geospatial-Intelligence Agency (NGA) has assumed a vital role by providing the FBI with geographic information from the U.S. Geological Survey and the military to bolster event security.

NGA, then known as the National Imagery and Mapping Agency (NIMA), helped officials in Salt Lake City survey the area around Olympic venues for vulnerabilities. The NGA routinely works with the FBI on large-scale events—from the Major League Baseball All-Star game to the annual September 11 commemoration ceremony in New York—susceptible to terrorist attacks.

Exercising on a large scale

The homeland security mission includes preparedness training through exercises that simulate terrorist attacks. The series of TOPOFF (shorthand for top officials) exercises mandated by Congress have been massive undertakings with logistical complexities rivaling an actual political, cultural, or sporting event. In April 2005, TOPOFF 3 in Connecticut and New Jersey simulated a bioterrorist attack and a car bombing. Participants found that various public agencies and the private sector had made progress building relationships, but there were lingering deficiencies in information management and coordination.

Operation Global Mirror, conducted in May 2004 in Colorado, was another simulated event that tested the Department of Homeland Security's National Incident Management System (NIMS) in the context of radioactive material released from an airplane. NIMS integrates emergency preparedness and response practices into a national framework for dealing with natural or terrorist events. The training exercise involved more than fifty federal, state, regional, and local emergency service agencies. It also was a test of military and civilian cooperation, and GIS helped bridge the gap by providing a common operating picture for incident managers.

Planning scenarios

To underscore the importance of preparedness exercises, the White House Homeland Security Council, the Department of Homeland Security, and state and local homeland security agencies developed fifteen all-hazards planning scenarios, covering nuclear, biological, chemical, explosive, and cyber attacks along with natural disasters. Each scenario includes a step-by-step checklist of tasks under the categories of awareness, preparedness, prevention, response, and recovery.

Homeland security exercises help shape the preparedness program for actual events with such routine safeguards as accounting for hazardous materials and explosive substances, and identifying sites for potential evacuation centers, incident command posts, and staging areas.

Whether dealing with simulated or potential terrorist events, GIS provides a platform for displaying venue locations, measuring consequences, and providing incident commanders the tools to deploy public safety resources to save lives and property.

CASE STUDY

Salt Lake City Winter Olympics

Five months after the September 11 terrorist attacks, the 2002 Winter Olympics opened in Salt Lake City, Utah, under security tighter than any other previous national sporting event. The event tested multiagency coordination using high-tech communications, and GIS played a pivotal role.

A satellite view of Salt Lake City with various venues of 2002 Winter Olympics marked with pushpins.

Source: National Aeronautics and Space Administration.

Athletes gathered in the Olympic spirit of competition and sportsmanship while the world watched a nation, still staggering from the destruction in New York City and Washington, D.C., show its resilience. Behind the scenes, federal, state, and local personnel were committed to keeping twenty venues safe within the region surrounding Salt Lake City. The single network event management program running at the Utah Olympic Public Safety Command (UOPSC) provided geospatial awareness through the use of ArcIMS and ArcView software. The GIS provided consistent, accurate, relevant, and timely data for those people monitoring the situation at the Olympic Command Center.

Enhancing situational awareness

More than three thousand users were connected to the system during the seventeen-day Olympics, enabling public safety agencies to prepare for and respond to emergencies. GIS tools allowed emergency and event management personnel to share critical information required for effective situational awareness. GIS experts from public safety and emergency medical agencies worked nonstop at the command center, security centers at each Olympic venue, and remote federal centers across the country.

Security personnel totaled more than ten thousand for an event that averaged seventy thousand visitors per day and the security budget surpassed $300 million. Most security measures were in place before September 11, but the attacks resulted in more federal funding for extra deterrents, including geospatial functions. Lidar data, gathered shortly before the games, was distributed to government agencies and private organizations to improve highly accurate mapping and 3D imagery.

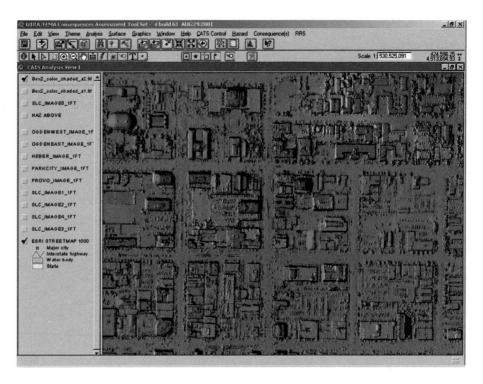

Lidar data contributed to highly accurate mapping and 3D imagery such as this map that relates color to height.

Source: ESRI and Science Applications International Corporation.

Digital orthoquarter quads of the region were updated and a helicopter was used to map radioactivity levels within much of the Salt Lake City region. That information became the base value to determine if any new "hot spots" indicating nuclear materials appeared during the games.

Transportation department chips in

The UOPSC was the key coordinating entity for more than sixty-five local, state, federal, and private organizations responsible for logistics and information. GIS experts created maps and data layers for the UOPSC. The Salt Lake City Division of Transportation contributed equipment, software, and databases the local agency already was using in day-to-day operations. The transportation division's extensive database of Salt Lake City orthoimagery, street maps, and building locations proved invaluable to the UOPSC for determining road closures and buffer zones in case of explosions. Transportation officials also printed more than one hundred maps to assist the Salt Lake Organizing Committee and governments.

The goal of all GIS participants was to create a dynamic common operating picture so that everyone would see the same information and deploy resources wisely. An Internet-based, dynamic mapping application proved to be the right tool for the job. The UOPSC turned to Science Applications International Corporation (SAIC), a San Diego research and

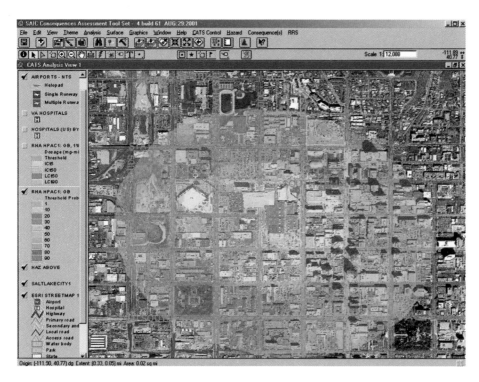

This map shows a simulated release of the nerve agent sarin at the Delta Center, site of the Olympic figure skating competition in Salt Lake City. The contours are showing the potential onset of symptoms.

Source: ESRI. Data courtesy of TANA/GDT, American Hospital Association, National Park Service, U.S. Census, and USGS.

engineering firm, and the Irvine-based E Team, an incident management software specialist, to create an information system that would be easy to use. The E Team software (a Web-based application) and SAIC's Consequences Assessment Tool Set (CATS) run in tandem and are integrated with ESRI software.

UOPSC members could access maps and data and communicate with each other over a secure, password protected environment. Users could review all current incident reports—traffic jams, weather emergencies, evacuations, and bomb threats—and sort through them by agency involved, type of incident, location, or other criteria. E Team also featured a scalable ArcIMS-enabled mapping system that used color-coded icons to display current incidents and help traffic and fire officials handle emergencies.

CATS, enhanced with ArcView and its Spatial Analyst and 3D Analyst extensions, provided tools to generate predictive models and perform casualty and damage assessments. Users could perform what-if scenarios for earthquakes, catastrophic weather events, water line contamination, explosions, nuclear incidents, or biochemical hazards. CATS was conceived for military uses, such as modeling explosions and natural hazards, but it expanded for use in state and local emergency management.

Data quality assured

More than one thousand different databases were used in the planning and consequence management of the Olympic events. The biggest obstacles were determining what data was relevant to emergency planning and standardizing the dataset format. A GIS working group, made up of representatives from data-contributing governments and commercial entities and Utah's homeland security team, took charge of assuring that data was current, accurate, and relevant. The availability of real-time data from remote sources enabled officials to adjust response plans to fit current conditions.

CompassCom of Denver provided UOPSC personnel with a software package to track more than five hundred athlete transport shuttles and a fleet of emergency vehicles. The package, combined with ArcView and MapObjects software, mapped bus routes and buffers around those routes to monitor vehicles, which were equipped with GPS units. Any deviation from a route would trigger an alert and drivers also would notify the command center if a problem arose. A GIS analysis, using lidar data and digital elevation modeling, determined the location of view sheds equipped with traffic and security cameras and the best placement for law enforcement snipers.

The massive public safety effort, perhaps the largest in U.S. history, required extraordinary coordination and cooperation from government at all levels. While the seventeen days of vigilance underscored the need for improved data standardization, the teamwork served as a model for regional homeland security. GIS helped bridge jurisdictional barriers, resulting in Winter Olympic Games that were memorable because of peak performances and a figure skating judging scandal, not the loss of human lives or destruction of property.

CASE STUDY

2004 Democratic National Convention

As delegates gathered in Boston to nominate Senator John Kerry as the Democratic Party's presidential candidate, GIS professionals were among those who worked feverishly behind the scenes to safeguard what homeland security officials designated a National Special Security Event. The Secret Service took charge of security measures, but it required close collaboration by a multitude of agencies to ensure the success of the Democratic National Convention (DNC) in July 2004.

A GIS working group at the convention was formed to coordinate geospatial issues, improve communication, and address specific GIS problems. Its focus was on data development and sharing, unified modeling, and mapping support. Dealing with disparate and often outdated or inaccurate data sources from various governmental agencies was a major challenge. The group, representing more than twenty federal, state, and local agencies, and private organizations, created a common dataset that all GIS personnel working the convention could access. The data was pulled from several sources, checked for accuracy and content, and distributed to users.

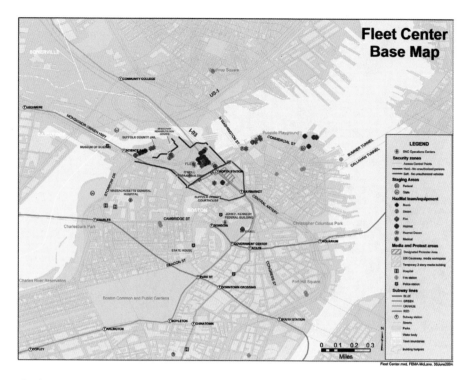

The Fleet Center Base Map, one of five basemaps produced.

Source: Lauren McLane, Federal Emergency Management Agency.

The group began meeting six months before the convention and produced an operations document to guide the collaborative process. The document spelled out standards for data format and type, naming conventions, data and map sharing, documentation, modeling group procedures, and contact information. The group also established procedures for distributing data and maps during the convention through a U.S. Environmental Protection Agency (EPA) file transfer protocol site, and for sharing updated information through an EPA message board.

Datasets meet needs

Besides existing data provided by various agencies, such as the Boston Police Department and EPA, the GIS working group developed new datasets tailored to the needs of the convention. This information included DNC event venues, location of medical and hazardous materials response teams, security zones, staging areas, transportation routes and road closures, and emergency operations centers. The Fleet Center, the arena where the convention was held, was close to the construction site of the Central Artery/Tunnel Project, commonly known as "The Big Dig," which had changed the local landscape dramatically. So, aerial orthophotos of the vicinity were taken in the spring of 2004.

Orthophoto of
the Fleet Center,
site of the 2004
Democratic
National
Convention.

Source: Lauren McLane,
Federal Emergency
Management Agency.

Five basemaps published in the field operations guide for the Federal Emergency
Management Agency ensured that all of the operations centers were viewing the same
information as the basis for their decisions. Thirty other maps showing critical facilities,
demographics, security areas, and DNC event-specific information were distributed to
members of the DNC Consequence Management Subcommittee. Because of the sensitive
nature of some information, such as the location of field response teams, the distribution of
data and maps was restricted until two weeks after the convention ended.

Modeling hazardous releases

A modeling group that included representatives from the EPA, local police department,
the National Oceanic and Atmospheric Administration (NOAA), and the National
Atmospheric Release Advisory Center (NARAC) convened at Boston police headquarters
during the convention to study the potential impacts of hazardous materials released into
the atmosphere. Their primary air dispersion models were Aerial Locations of Hazardous
Atmospheres, Hazard Prediction and Assessment Capability, and NARAC's Web-based or
reach-back air model.

The modeling group provided emergency responders at the event with a consistent
plume model. Boston police, in consultation with NOAA, positioned meteorological
equipment throughout the city to measure wind force and upper level atmospheric weather.
All meteorological data for the Boston area was consolidated and distributed through the
Defense Threat Reduction Agency, an office of the U.S. Department of Defense.

Police officials also used a GIS application to collect information on the Fleet Center and other DNC facilities along with locations that housed important attendees. The building information was available for emergency response and incident management. After the convention, the Boston Police Department used Tablet PCs and ArcEditor to inventory hotels, municipal structures, schools, museums, and other significant city buildings, and developed an incident plan for each building.

The DNC went off without a serious incident and the collaborative effort of the GIS working group was deemed successful. The process confirmed that it is essential for multiple agencies joined for a common purpose to work with one set of data and maps and consistent models.

CASE STUDY

Super Bowl XLI

Teamwork enabled the Indianapolis Colts to win Super Bowl XLI on February 4, 2007. Teamwork also protected the venue, Dolphin Stadium, and the South Florida region from threats.

Coordinating multiple agencies and integrating the technology to allow those agencies to communicate with each other required extensive planning. That no major incidents occurred before, during, or after the football game within the Miami-Dade County region was a testament to the hard work behind the scenes, which included a strong GIS presence.

Merging software

The Miami-Dade County Office of Emergency Management and Homeland Security (OEM&HS), as the local lead agency, faced the challenge of merging different software and data into a compatible platform for easy access on an intranet Web application. The OEM&HS was well positioned to meet the challenge, having relied on GIS to coordinate responses to hurricanes, fires, hazardous material spills, and other emergency situations. OEM&HS also had an established track record of assessing vulnerabilities and maintaining current, relevant data.

Soheila Ajabshir, GIS coordinator for the OEM&HS, said the U.S. National Grid (USNG) referencing system for the region was foundational for aiding the Urban Search and Rescue (USAR) team. CAD drawings of Dolphin Stadium and key perimeter sites, digital orthophotos, ESRI and Oracle data, were combined with the grid system to create a common operating picture.

All countywide critical facilities such as hospitals, schools, and transportation facilities were mapped or printed for different agencies prior to and during the week leading up to the Super Bowl. Emergency managers and security personnel at the Miami-Dade County Emergency Operations Center (EOC) monitored traffic, airspace, waterways, and crowds while investigating alerts, tips, and any unusual events.

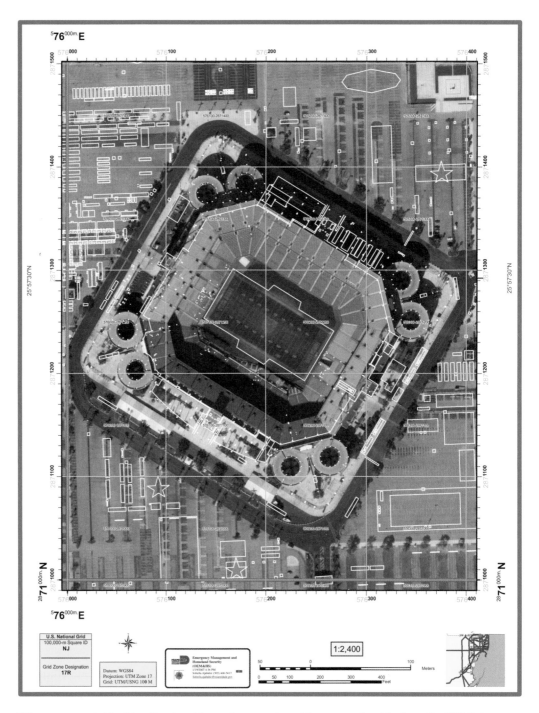

This map was created for Urban Search and Rescue team members at the command post for Super Bowl XLI. It shows U.S. National Grids overlaying Dolphin Stadium aerial photograph. The CAD annotation is not shown for the purpose of security.

Source: Soheila Ajabshir, Miami-Dade County.

Monitoring pregame events

Pre-Super Bowl events, such as major parties and television and radio tapings, were monitored at the EOC over the several days before the game itself. During the game, Ajabshir's eyes were glued on computer screens, not the Super Bowl coverage.

Every agency at the EOC, including representatives from counties, municipalities, nonprofit organizations, and state and federal agencies at the EOC were linked via E Team software, NC4's situational response solution. NC4 (National Center for Crisis and Continuity Coordination) is based in El Segundo, California, and provides real-time alerts and communication on significant incidents affecting business and government.

GIS and E Team enable multiple organizations involved in preparing for and responding to emergencies to collaborate and manage their efforts from a single common view and coordination point. The geospatial technology at the Super Bowl delivered key information to the right people at the right time, regardless of their location. Officials could observe, assess, and analyze data from spontaneous and planned events to make accurate and timely decisions.

This map was created for Miami-Dade EOC, showing Miami Beach area parties and events scheduled for February 1, three days before Super Bowl XLI.

Source: Soheila Ajabshir, Miami-Dade County.

At the EOC in Miami, users were able to post and retrieve information on the intranet site Miami-Dade Viewer to evaluate current situations and potential incidents. As reports, graphics, maps, or satellite displays were updated, the new information was automatically available to everyone.

ArcGIS server and E Team's program and software combined to allow users to access maps and conduct GIS analysis using a simple browser. Users could conduct incident modeling, access incident data, and obtain situational awareness with highly detailed geographic data. This data was dynamically updated and provided to all users with current map displays, easily and quickly.

GIS applications were used to map USNG in combination with transportation, critical facilities, digital ortho images, and live data from E Team's planned event in order to exhibit and monitor a common picture at the EOC and the command post near the stadium.

From a threat standpoint, the big event proved uneventful, which suited Ajabshir and others at the EOC just fine.

References

ArcNews. 2002. GIS works behind the scenes to assist in emergency preparedness; Salt Lake Olympics set new safety standard. Summer. http://www.esri.com/news/arcnews/summer02articles/salt-lake-olympics.html.

Barnes, Scottie. 2004. Salt Lake City's Olympic effort. *GeoIntelligence* (January 1).

Davis, David B. 2002. GIS and the Salt Lake 2002 Winter Olympics. *GISVision* (April). http://www.gisvisionmag.com/vision.php?article=200204/special.html.

E Team Web site. Delivering the gold: E Team and SAIC create the Olympic Command Center for Security Training and Oversight of the Winter Games. http://www.eteam.com/resources/salt_lake_winter_olympics_web.pdf.

———. NC4's national incident monitoring centers and incident management technology at the heart of Super Bowl security. http://www.eteam.com/press/releases/news_superbowl07.html.

McLane, Lauren, and Johanna Meyer. 2005. GIS for a homeland security event: The Democratic National Convention. Presentation at the twenty-fifth annual ESRI International User Conference, San Diego, Calif., July 25–29. http://gis.esri.com/library/userconf/proc05/papers/pap1579.pdf.

Price, Mike. 2004. Incident command and GIS. *ArcUser Online* (October–December). http://www.esri.com/news/arcuser/1104/nims1of2.html.

Public Technology Institute. 2006. Using GIS to support emergency management for homeland security. December.

U.S. Department of Homeland Security. 2004. The Department of Homeland Security partners with state and locals to protect Democratic National Convention. July 14. http://www.gwu.edu/~action/2004/demconv04/securityfact071404.html.

U.S. Department of Transportation. 2005. Transportation Management and Security during the 2004 Democratic National Convention. January 5. http://www.itsdocs.fhwa.dot.gov/JPODOCS/REPTS_TE/14120.htm.

6
Looking ahead

Just as everyone knows that exercise and a proper diet are essential to good health, few would deny that geospatial data is crucial to homeland security. But eating properly and feeding a comprehensive geodatabase both take resolve. Under the homeland security umbrella, government and the private sector are striving for consistency, coordination, and collaboration through GIS. The pace, however, is less than brisk.

GIS needs consistent funding at all levels of government if it is to reach its full potential as an integrating mechanism to support and enhance homeland security effectiveness. As the preceding case studies illustrated, an investment in GIS meets the homeland security imperative while paying dividends in a variety of routine local government functions, such as enhancing operations and maintenance, preparing for emergencies, and fighting crime. So far, the federal government has provided the bulk of the seed money for state and local linkages to national GIS initiatives, but the long-range goal is for local, state, and regional homeland security programs that connect to a national GIS to become self-sustaining.

Federal initiatives

The federal government, through its various initiatives, has set a firm foundation for GIS technology. The National Spatial Data Infrastructure (NSDI) embodies the technology, policies, and personnel working to promote data sharing among governments, academia, business, and nonprofit organizations. The NSDI seeks to reduce duplication of effort among agencies, improve quality, and reduce costs related to geographic information, and make geographic data more accessible and beneficial to the public.

The Federal Geographic Data Committee (FGDC) champions the NSDI by encouraging the coordinated development, use, sharing, and dissemination of geospatial data on a national basis. The FGDC membership consists of representatives from nineteen federal agencies, including Cabinet-level departments and high-profile agencies such as the Environmental Protection Agency and the Federal Emergency Management Agency.

A breakthrough program that evolved from the NSDI initiative was the Department of the Interior's Geospatial One-Stop Web portal, www.geodata.gov, for sharing federal, state, local, tribal, and private geographic information. This online tool proved its worth as a critical resource for maps and data during the response to Hurricanes Katrina and Rita in 2005. ESRI, contracting with Geospatial One-Stop, helped deliver more than fifty types of information to nearly one hundred users through the password-protected GIS for the Gulf site.

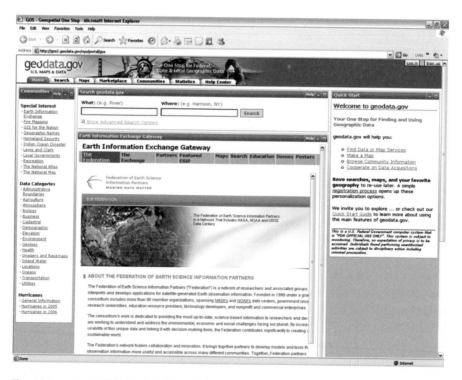

The U.S. Department of the Interior's Geospatial One-Stop Web portal, www.geodata.gov, enables federal, state, local, tribal, and private entities to share geographic information. This online tool has been especially valuable during national emergencies such as Hurricane Katrina.

Screenshot courtesy of geodata.gov U.S. Maps & Data, 2007.

The National Map

A key component of the NSDI, the National Map, is replacing outdated paper basemaps with digital geographic base information that can be updated weekly. It represents a profound change for mapping in the United States by expanding the role of the U.S. Geological Survey (USGS) to include integrating digital and disparate geospatial information and facilitating partnerships for data sharing. The USGS science strategy for geographic research through 2015 foresees this scenario:

> By 2015, a host of consumer items will have integrated spatial data into their operations by taking advantage of wireless communications technology, mobile and portable computing devices, global position systems (GPS), and the National Map. The National Map will become the preferred gateway for popular access into geographical data for everyday tasks. For example, farmers may walk their fields looking at satellite imagery centered on their location, historians may browse historical air photos while walking through a nineteenth century industrial plant, and tourist buses may show a digital display of the day's sightseeing circuit that changes themes depending on the tour guide's discussion.[1]

For this scenario to come true, geographic information science (GIScience), the research behind GIS, must continually develop innovative ways to process and access increasing volumes of data, the USGS report acknowledged. And, the National Map will assume its role as an efficient framework for organizing and maintaining disparate databases only through strategic collaboration and partnerships facilitated chiefly by the USGS.

A 2004 study projected $2 billion in National Map benefits over its thirty-year lifespan due to data that is current, integrated, consistent, complete, and more accessible than in previous formats. The National Map ultimately will house a complete national set of digital topographic and orthophoto data, as well as other detailed layers of information of transportation, administrative boundaries, land-use, and land-cover information. This up-to-date national dataset available to state and local governments would further the goals of homeland security but not without challenges. Building a consistent, comprehensive geographic data repository requires government at all levels to upgrade their GIS data, which takes time and money.

From the bottom up

Stimulating GIS programs at the grassroots level is essential if a comprehensive national geospatial program for homeland security is to succeed. A 2004 survey found that state, local, and tribal interests were reluctant to participate in the National Map because it was unclear how the initiative would provide direct benefits. A report on the progress of the National Map in 2005 encouraged the National Geospatial Program Office (NGPO) of the USGS to tailor its programs to states and regions as a further incentive for participation in the ambitious initiative.

---

1. Gerald McMahon et.al. 2005. "Geography for a Changing World. A Science Strategy for the Geographic Research of the U.S. Geological Survey 2005–2015." (Reston, Va.: U.S. Department of the Interior, U.S. Geological Survey), 37.

The National Map represents a profound change for mapping in the United States by expanding the role of the U.S. Geological Survey (USGS) to include integrating digital and disparate geospatial information and facilitating partnerships for data sharing.

Screenshot courtesy of USGS, 2007.

The report stressed the importance of regional groups, aided by state and federal GIS liaisons, to develop consistent data and meet the needs of local partners. A federally funded pilot program in California, the Bay Area Regional GIS Council (BAR-GC), has proven the viability of this approach by knocking down administrative barriers to efficient data sharing. Similarly, the Ohio Geographically Referenced Information Program (OGRIP) has emerged as the primary coordinator of GIS initiatives involving federal, state, regional, and local governments.

The NGPO's GIS for the Nation data model, also known as the Geospatial Bluebook, identifies standards and best practices that have worked throughout the country and offers implementation templates for local communities. The ultimate goal is an interoperable national system built on current, relevant, high-quality data layers. The data model strives to address local and national needs for both focus and flexibility in GIS. This vision is based on the principle that data generally should be collected locally and shared broadly through regional data centers that state and national agencies can tap into. Building on the bluebook model, the Department of Homeland Security Geospatial Data Model (DHS GDM) was planned for phasing in through the 2007 fiscal year as an extract, transform, and load (ETL) template for content aggregation. (ETL is a tool used for migrating data and database formats.)

The USGS ten-year science strategy projection calls for developing innovative methods of modeling and information synthesis, fusion, and visualization to create "new knowledge" from geographic data. The report predicts that by 2015:

> USGS scientists will routinely provide global models of areas, populations, and resources of any affected world locally. These models will be instrumental in the efforts to mitigate loss of life and property through preparedness, prevention, and response. Other USGS models that use the same massive datasets and fusion methods will provide risk and hazards assessment modeling for preparedness and prevention of homeland security threats.[2]

In June 2006, the Department of Homeland Security reiterated the importance of information sharing through its National Infrastructure Protection Plan. The plan identified priority actions—such as promoting public awareness and encouraging new technology—toward building an effective, efficient protection program over the long term. Still, old habits die hard and the Department of Homeland Security has been slow to remove its own obstacles to data sharing, primarily by insisting that much information remain classified. Officials with the Urban and Regional Information Systems Association (URISA), a leading association of GIS professionals that strives for greater spatial data integration in government, contend that while security is essential, it also must be reasonable if the goal of widely shared standardized maps and data is to be reached.

Terrorism under study

In the academic world, homeland security increasingly is a focus of research and technology at several institutions. The CREATE Homeland Security Center at the University of Southern California was the first of six think tanks funded by the federal government to predict the impact of terrorist events, estimate their economic consequences, and identify where the country is most vulnerable.

CREATE (Center for Risk and Economic Analysis of Terrorism Events) was initially funded in 2003 with a $12 million, three-year grant from the U.S. Department of Homeland Security. GIS supports CREATE's technological output with maps, spatial datasets, components, and applications. The ArcGIS platform helps researchers manage data and maps and develop models and software applications. In the area of emergency response, CREATE has developed GIS-based tools to help response agencies allocate and deploy personnel and equipment more efficiently. The research also has involved building models integrated into a GIS for distributing vaccines, medicine, and other inventories.

The center sees a long-term investment in economic modeling and analysis, with an emphasis on integrating risk and economic assessment. CREATE also planned to expand the center's scope to include nonterrorist disasters and to depend less on federal DHS funding.

Clearly, the nation's security hinges on the ability to create, maintain, and distribute accurate and comprehensive data. Because geographic location is a key feature of about 90 percent of government data, GIS will remain at the forefront when dealing with the physical, psychological, and economic consequences of natural disasters or terrorist attacks.

2. Ibid., 41.

References

CREATE (University of Southern California Center for Risk and Economic Analysis of Terrorism Events). 2006. *CREATE Annual Report Year Two, March 2004 to March 2006.* March 31. http://www.usc.edu/dept/create/reports/CREATE-SemiAnnualONR-Final%20Year%202.pdf.

Essex, David. (2006) DHS special report. FEMA maps out a better response. *Government Computer News,* June 19. http://www.gcn.com/print/25_16/41081-1.html.

Federal Geographic Data Committee Web site. http://www.fgdc.gov/.

Gordon, Larry. 2006. Pondering the costs of terror protection. A USC think tank takes a multi-faceted approach in assessing the risks of attack and determining the best defense. *Los Angeles Times.* July 10. http://www.latimes.com/news/local/la-me-risk10jul10,1,39020.story?coll=la-headlines-california.

Halsing, David, Kevin Theissen, and Richard Bernknopf. 2004. A cost-benefit analysis of the National Map. U.S. Department of the Interior, U.S. Geological Survey. May. http://pubs.usgs.gov/circ/2004/1271/c1271.pdf.

Imwalle, Shane, and Jime Kiles. 2003. Leveraging GIS data creation and maintenance for homeland security, emergency management applications. *The Military Engineer* (July–August). http://www.woolpert.com/asp/articles/Homeland%20Security.asp.

McMahon, Gerald, et. al. 2005. Geography for a changing world. A science strategy for the geographic research of the U.S. Geological Survey, 2005–2015. U.S. Department of the Interior, U.S. Geological Survey. Report Number 1281. http://geography.usgs.gov/documents/USGSGeographySciencePlan.pdf.

Schwartz, Karen. 2003. Mapping a more secure future. *GovExec* (February 15). http://www.govexec.com/features/0203/0203managetech1.htm.

U.S. Department of Homeland Security. 2006. *National Infrastructure Protection Plan.* June. http://www.dhs.gov/interweb/assetlibrary/NIPP_Plan.pdf.

U.S. Department of the Interior, U.S. Geological Survey. 2005. Final Report. The National Map Partnership Project. http://nationalmap.gov/report/NSGIC_TNM_Report_102705_V6.doc.

_____. 2005. Geospatial One-Stop project awards portal contract. Press release. January. http://www.doi.gov/news/05_News_Releases/050131c.

Acronyms and abbreviations

ACTIC   Arizona Counter Terrorism Information Center. Coordinates information sharing among law enforcement authorities to deter terrorist activities.

AIC   Architecture and Infrastructure Committee. Key support entity of the federal Chief Information Officers Council.

ALOHA   Aerial Location of Hazardous Atmospheres program. Computer program developed by the National Oceanographic and Atmospheric Administration for modeling plumes of chemical releases.

ARES   Amateur Radio Emergency Services. A public service organization of licensed amateur radio operators who have voluntarily registered their qualifications and equipment to provide emergency communications for public service events as needed.

ASI   American Shield Initiative. Border patrol program under DHS that provides full nationwide border coverage, including cameras, ground sensors, unmanned aerial vehicles (UAVs), and other technologies, to monitor the border and provide common information awareness of areas between border-crossing points.

BAR-GC   Bay Area Regional GIS Council (San Francisco). Partners in ESRI pilot project for homeland security data-sharing approach.

BRAC   Bioterrorism Response Advisory Committee. A committee consisting of the Department of Health partners and stakeholders that advise the Department of Health on its plan for bioterrorism preparedness and response.

CARVER   Criticality, Accessibility, Recuperability, Vulnerability, Effect, Recognizability. CARVER is a target-selection process or matrix.

CATIC   California Anti-Terrorism Information Center. California's clearinghouse for investigating terrorist related activities.

CATS   Consequences Assessment Tool Set. Analysis system created by Science Applications International Corporation to prepare for and manage disasters.

CDC   Centers for Disease Control and Prevention. An agency of the U.S. Department of Health and Human Services based in Atlanta, Georgia, that works to protect public health and the safety of people.

CEPPO   Chemical Emergency Preparedness and Prevention Office. An entity of the Environmental Protection Agency that offers technical assistance and education about chemical hazards.

CERT   Citizen Emergency Response Team. An organization chartered by an information system owner to coordinate or accomplish necessary actions response to computer emergency incidents that threaten the availability or integrity of its information systems.

CIGT   Center for Innovative Geospatial Technology. Organizational unit related to the ESRI Professional Services Department.

CONOPS	Concept of operations. A type of white paper that shows how information systems work, how they can be improved, and what the impacts are for system development.
CONPLAN	U.S. Government Interagency Domestic Terrorism Concept of Operations Plan. The CONPLAN was designed to provide overall guidance to federal, state, and local agencies concerning how the federal government would respond to a potential or actual terrorist threat or incident, particularly one involving weapons of mass destruction (WMD). Six federal agencies are signatories to the plan: Department of Justice, Federal Emergency Management Agency, Environmental Protection Agency, Department of Energy, Department of Defense, and the Department of Health and Human Services. The CONPLAN was created to implement Presidential Decision Directive 39 (1995), which sets forth U.S. policy on counterterrorism.
CREATE	Center for Risk and Economic Analysis of Terrorism Events. Located at the University of Southern California and funded by the Department of Homeland Security, the center evaluates the risks, costs, and consquences of terrorism.
CSEPP	Chemical Stockpile Emergency Preparedness Program. One facet of the multihazard readiness program in eight U.S. communities dealing with natural and man-made emergencies. Depending on the location of the community, such emergencies may include tornadoes, hurricanes, earthquakes, floods, fires, hazardous materials spills or releases, and transportation and industrial accidents. The program's goal is to improve preparedness to protect the people of these communities in the unlikely event of an accident involving this country's stockpiles of obsolete chemical munitions.
CSI	Container Security Initiative. Designed to help protect the United States and a large portion of the global trading system from terrorists who might use container transport to hide weapons of mass destruction and related materials without disrupting legitimate flow of cargo. CSI requires bilateral agreements be created with other governments to target and prescreen high-risk containers in overseas seaports before they are shipped to the United States. Customs inspectors (prescreeners) will also be stationed in CSI ports to work with their overseas counterparts.
CTM	Cooperative Topographic Mapping Program. A U.S. Geological Survey program that provides the nation with access to current, accurate, and consistent base geographic data and derivative products such as topographic maps. The program focuses on partnerships for data sharing that can be used as content for the National Map. Within CTM there are seven framework themes: orthoimagery, elevation, transportation, hydrography, boundaries, structures, and names.
DCMI	Dublin Core Metadata Initiative. An organization dedicated to promoting the widespread adoption of interoperable metadata standards and developing specialized metadata vocabularies for describing resources that enable more intelligent information discovery systems.
DHS	U.S Department of Homeland Security. A consolidation of twenty-two government agencies to provide the unifying core for the vast national network of organizations and institutions involved in efforts to secure the nation. (Note: Many state governments use "DHS" to refer to their state Department of Human Services and Department of Health Services.)

DHS GDM   Department of Homeland Security Geospatial Data Model. A standards-based, logical data model used for collection, discovery, storage, and sharing of homeland security geospatial data.

DIA   Defense Intelligence Agency. The DIA mission involves combat support and being a producer and manager of foreign military intelligence.

DMA   Disaster Mitigation Act. A U.S. legislative act passed in October 2000 that requires state and local coordination for disaster preparedness.

DMAT   Disaster Medical Assistance Team. A DMAT is a deployable national asset that can provide triage, medical, or surgical stabilization and continued monitoring and care of patients until they can be evacuated to locations where they will receive definitive medical care. Specialty DMATs can also be deployed to address mass burn injuries, pediatric care requirements, chemical injury or contamination, and so forth. The DMAT program is managed by the Department of Homeland Security in coordination with the Department of Health and Human Services.

DMORT   Disaster Mortuary Operational Response Team. A DMORT is a deployable national asset that can assist local authorities in providing victim identification and mortuary services, including temporary morgue facilities. Victims can be identified by fingerprint, forensic dental, and/or forensic pathology/anthropology methods. Mortuary services include processing, preparation, and disposition of remains. The DMORT program is managed by the Department of Homeland Security in coordination with the Department of Health and Human Services.

DMSO   Defense Modeling and Simulation Office. The catalyst organization for Department of Defense modeling and simulation ensuring that M&S technology development is consistent with other related initiatives.

DoD   U.S. Department of Defense.

DOI   U.S. Department of the Interior.

DTRA   Defense Threat Reduction Agency. DTRA's mission is to safeguard America and its friends from weapons of mass destruction by reducing the present threat and preparing for future threats.

EAS   Emergency Alert System. Established by the FCC in November of 1994 to replace the Emergency Broadcast System (EBS) as a tool the president and others might use to warn the public about emergency situations.

EMAC   Emergency Management Assistance Compact. A legally binding mutual aid agreement and partnership among the states that allows them to assist one another during emergencies and disasters.

EMMA   Emergency Management Mapping Application. Web-based software that integrates disparate databases for emergency management applications.

EPA   U.S. Environmental Protection Agency.

EPCRA   Emergency Planning and Community Right-to-Know Act. Federal Emergency Management Agency (FEMA) executive mandate to share information with local communities.

FAA   Federal Aviation Administration. An agency of the U.S. Department of Transportation responsible for the safety of civil aviation.

FAST In the United States and Canada, Free and Secure Trade. FAST is a harmonized clearance process for shipments of known compliant importers. FAST is for shipments destined to the United States (from Canada or Mexico) transported by highway. For trucks to use FAST lane processing, the Mexican manufacturer must be C-TPAT approved, the U.S. importer of record must be C-certified, and the commercial driver must possess a valid FAST Commercial License. Cargo release methods for FAST shipments are the National Customs Automated Prototype (NCAP) and the Prearrival Processing System (PAPS).

FEA Federal Enterprise Architecture. Part of the E-Government initiative and a business-based framework for government-wide improvement.

FEMA Federal Emergency Management Agency. An agency in the Department of Homeland Security. FEMA's mission is to lead the effort to prepare the nation for all hazards and effectively manage federal response and recovery efforts following any national incident. FEMA also initiates proactive mitigation activities, trains first responders, and manages the National Flood Insurance Program and the U.S. Fire Administration.

FGDC The Federal Geographic Data Committee. The FGDC champions the National Spatial Data Infrastructure by encouraging the coordinated development, use, sharing, and dissemination of geospatial data on a national basis.

FinCEN Financial Crimes Enforcement Network. A U.S. Department of Treasury program established in 1990 to implement and oversee policies related to money laundering. Provides information sharing and strategic analysis of domestic and worldwide money laundering developments, trends, and patterns.

FRERP Federal Radiological Emergency Response Plan. The plan that describes the federal response to the radiological and on-site technical aspects of an emergency in the United States and identifies the lead federal agency for the event. The events include one involving the Nuclear Regulatory Commission or state licensee, the U.S. Department of Energy or the U.S. Department of Defense property, a space launch, an occurrence outside the United States but affecting the United States, and one involving radium or accelerator-produced material. Transportation events are those involving the U.S. Nuclear Regulatory Commission, state licensee, U.S. Department of Energy, or U.S. Department of Defense.

FRP Federal Response Plan. The plan designed to address the consequences of any disaster or emergency situation in which there is a need for federal assistance under the authorities of the Stafford Act. Twenty-seven federal departments and agencies, including the American Red Cross, are signatories to the plan.

FWS U.S. Fish and Wildlife Service.

GEODE Geologic discipline contributions to the NSDI Web application for data viewing related to geologic and natural hazards. GEODE has recently been adopted as the data delivery system of the geologic discipline and continues to provide unbiased scientific and energy-related data to the public via a fully functional, Web-accessed GIS map server.

GIDI Geospatial intelligence database integration. GIDI is a GIS enterprise that integrates existing NGA databases and production tools.

GIG Global information grid. National Security Agency concept for a seamless, secure, interconnected, omnipresent information network.

GOS	Global Observing System (European). A system to provide from-the-earth and from-outer-space observations of the state of the atmosphere and ocean surface for the preparation of weather analyses, forecasts, advisories, and warnings for climate monitoring and environmental activities carried out under program of WMO and other relevant international organizations.
HAZUS-MH	Hazard United States-Multi-Hazard. Refers to a FEMA risk-assessment software program for analyzing potential losses from floods, hurricane winds, and earthquakes. In HAZUS-MH, current scientific and engineering knowledge is coupled with ESRI ArcGIS technology to produce estimates of hazard-related damage before or after a disaster occurs.
HEICS	Hospital Emergency Incident Command System. The Incident Command System framework specific to hospitals, developed by the State of California, and used by many hospitals. It specifies the chain of command and functional positions that may be required during a hospital's response to an emergency.
HMGP	Hazard Mitigation Grant Program. FEMA program to provide grants that reduce the loss of life and property due to natural disasters and to enable mitigation measures to be implemented during the immediate recovery from a disaster.
HSEEP	Homeland Security Exercise and Evaluation Program. HSEEP is a threat-and performance-based exercise program that includes a cycle, mix, and range of exercise activities of varying degrees of complexity and interaction.
HSIP	Homeland Security Infrastructure Protection. A database model (schema) for infrastructure.
HSOC	Homeland Security Operations Center. HSOC serves as the primary national-level hub for operational communications and information pertaining to domestic incident management. Located at DHS headquarters, HSOC provides threat monitoring and situational awareness for domestic incident management on a 24/7 basis.
HSPD-5	Homeland Security Presidential Directive 5. This directive enhances the ability of the United States to manage domestic incidents by establishing a single, comprehensive national incident management system.
HSPD-7	Homeland Security Presidential Directive 7. This directive establishes a national policy for federal departments and agencies to identify and prioritize United States critical infrastructure and key resources and to protect them from terrorist attacks.
HSPD-8	Homeland Security Presidential Directive 8. This directive establishes policies to strengthen the preparedness of the United States to prevent and respond to threatened or actual domestic terrorist attacks, major disasters, and other emergencies by requiring a national domestic all-hazards preparedness goal, establishing mechanisms for improved delivery of federal preparedness assistance to state and local governments, and outlining actions to strengthen preparedness capabilities of federal, state, and local entities.
HSPD-9	Homeland Security Presidential Directive 9. This directive establishes a national policy to defend the agriculture and food system against terrorist attacks, major disasters, and other emergencies.
HSPD-10	Homeland Security Presidential Directive 10. This directive sets the framework for the nation's biodefenses.

IAEA   International Atomic Energy Agency. Under the United Nations, IAEA serves as the world's center of nuclear cooperation and standards and tracking of nuclear materials at more than nine hundred civilian facilities worldwide to verify their peaceful use.

IAIP   Information Analysis and Infrastructure Protection. IAIP serves as a national critical infrastructure threat assessment, warning, and vulnerability entity.

IIMG   Interagency Incident Management Group. The IIMG is made up of senior representatives from federal departments and agencies, nongovernmental organizations, and DHS components to facilitate national-level situation awareness, policy coordination, and incident coordination.

ISO   International Organization for Standardization. An international standard-setting body composed of representatives from various national standards bodies. Founded in 1947, the organization produces worldwide industrial and commercial standards.

LEPC   Local Emergency Planning Committee. A term used in the Emergency Planning and Community Right-to-Know Act (42 U.S.C. 11001.1986). EPCRA, also known as Title II of the Superfund Amendments and Reauthorization Act, was enacted by Congress as the national legislation on community safety. It was designed to help local communities protect public health, safety, and the environment from chemical hazards. To implement EPCRA, Congress required each state to appoint a State Emergency Response Commission and required each SERC to divide its state into emergency planning districts and name an LEPC for each district. Board representation by firefighters, hazardous materials specialists, health officials, government and media representatives, community groups, industrial facilities, and emergency managers helps ensure that all the necessary perspectives are represented on the LEPC.

MAG-C   Mission Assurance Governance Committee. A federal interagency working group whose mission is to share information and promulgate best practices on mission assurance to the diverse community of practitioners dedicated to ensuring the mission-essential functions of the federal government.

MIPT   Memorial Institute for Prevention of Terrorism. A nonprofit organization in Oklahoma City that sponsors research to discover equipment, training, and procedures to help prevent terrorism and respond to it.

NDMS   National Disaster Medical System. A cooperative, asset-sharing partnership between the Department of Health and Human Services, the Department of Veterans Affairs, the Department of Homeland Security, and the Department of Defense. NDMS provides resources for meeting the continuity of care and mental health services requirements of Emergency Support Function 8 in the Federal Response Plan.

NED   National elevation dataset. A USGS product for digital elevation models. NED is designed to provide national elevation data in a seamless form with a consistent datum, elevation unit, and projection. Data corrections were made in the NED assembly process to minimize artifacts, permit edgematching, and fill sliver areas of missing data.

NGA   National Geospatial–Intelligence Agency. A federal agency responsible for the collection, analysis, and distribution of geospatial intelligence in support of national security. NGA was formerly known as the National Imagery and Mapping Agency (NIMA) and is part of the Department of Defense (DoD), but also has responsibilities to customers outside the DoD.

NGPO   National Geospatial Programs Office. USGS office intended to house the National Map, National Atlas, Geospatial One-Stop, and FGDC in a single program office.

NGTOC   National Geospatial Technical Operations Center. A USGS production operations group based in Rolla, Missouri, that consolidates five mapping centers, including the Rolla center, into one. Others are located in Virginia, South Dakota, Colorado, and California.

NHD   National hydrography dataset. A comprehensive set of digital spatial data that contains information about surface water features such as lakes, ponds, streams, rivers, springs, and wells.

NICC   National Interagency Coordination Center. Wildfire information and coordination Web site.

NIFC   National Interagency Fire Center. Based in Boise, Idaho, the NIFC is the nation's support center for wildland firefighting. Eight different agencies and organizations are part of NIFC. Decisions are made using the interagency cooperation concept because NIFC has no single director or manager.

NIIMS   National Interagency Incident Management System. Consists of five major subsystems that collectively provide a total systems approach to all-risk incident management. The subsystems are the Incident Command System (ICS), training, qualifications and certification, supporting technologies, and publication management. NIIMS is related to the Wildland Fire Program and is not to be confused with NIMS.

NIMS   National Incident Management System. A system mandated by HSPD-5 that provides a consistent nationwide approach for federal, state, local, and tribal governments, the private sector, and nongovernmental organizations to work effectively and efficiently together to prepare for, respond to, and recover from domestic incidents, regardless of cause, size, or complexity. To provide for interoperability and compatibility among federal, state, local, and tribal capabilities, NIMS includes a core set of concepts, principles, and terminology. HSPD-5 identifies these as ICS, multiagency coordination systems, training, identification and management of resources (including systems for classifying types of resources), qualification and certification, and the collection, tracking, and reporting of incident information and incident resources.

NRO   National Reconnaissance Office. The NRO designs, builds, and operates the nation's reconnaissance satellites.

NRP   National Response Plan. A plan mandated by HSPD-5 that integrates federal domestic prevention, preparedness, response, and recovery plans into one all-disciplines, all-hazards plan.

NSDI   National Spatial Data Infrastructure. The 1994 Presidential Executive Order #12906 mandated the sharing of geospatial information. Amended by Executive Order 13286 in 2003, it includes technology, policies, standards, and human resources necessary to acquire, process, store, distribute, and improve use of geospatial data.

NSGIC   National States Geographic Information Council. An organization committed to efficient and effective government through the prudent adoption of geospatial information technologies. Members include state GIS managers and coordinators along with representatives from federal agencies, local government, the private sector, academia, and other professional organizations.

NWIS   National Water Information System. A USGS data dissemination Web site.

ODP	Office for Domestic Preparedness. A Department of Homeland Security agency and the primary office responsible for providing training, funds for the purchase of equipment, support for the planning and execution of exercises, technical assistance, and other support to assist states and local jurisdictions to prevent, plan for, and respond to acts of terrorism.
OGC	Open Geospatial Consortium, Inc. A standards development group composed of approximately 274 corporate and academic and government GIS representatives.
OGRIP	Ohio Geographically Referenced Information Program. Coordinates the state's GIS initiatives at various governmental levels.
OWTS	One-Way Transfer System. Raytheon technology for migration of data between secure computer systems.
PCII	Protected Critical Infrastructure Information. Program to encourage data sharing between industry and government.
PSWN	Public Safety Wireless Network. Joint Department of Justice and Department of Treasury program that promotes federal and local government communications interoperability and sponsors innovative pilot projects.
TOPOFF	Top Officials. Congressionally mandated antiterrorism exercise.
UASI	Urban Area Security Initiative. DHS grant process for city homeland security.
UOPSC	Utah Olympic Public Safety Command. Winter Olympics coordinating entity.
US-CERT	United States Computer Emergency Readiness Team. Charged with protecting national Internet infrastructure and coordinating defense from cyberattack.
USFS	U.S. Forest Service.
USGS	U.S. Geological Survey.
WMO	World Meteorological Organization. There are many acronyms related to partners within the WMO; these are listed on its Web site: http://www.wmo .int/web/www/Earthwatch/Partner-wmo.html.

Sources

Compendium of Federal Terrorism Training for State and Local Audiences.
   http://www.dola.colorado.gov/dem/publications/FEMA_Training_Compendium.pdf.

NIMS Glossary of Key Terms for the Purposes of NIMS.
   http://www.nimsonline.com/nims_3_04/glossary_of_key_terms.htm.

FEMA. http://training.fema.gov/EMIWeb/IS/is14/glossary.htm.

Homeland Security Glossary of TermsNORTHCOM.
   http://www.dhs.gov/xlibrary/assets/foia/US-VISIT_RFIDattachB.pdf.

Joint Chiefs of Staff Publications. http://www.dtic.mil/doctrine/s_index.html.

National Mutual Aid & Resource Management Initiative Glossary of Terms and Definitions.
   http://www.fema.gov/pdf/emergency/nims/507_Mutual_Aid_Glossary.pdf.

Legal Information Institute, U.S. Code Library. http://www.law.cornell.edu/uscode/.

Glossary of homeland security terms

access controls Procedures and controls that limit or detect access to minimum essential infrastructure resource elements (people, technology, applications, data, facilities), thereby protecting these resources against loss of integrity, confidentiality, accountability, and availability.

access control system elements Detection measures used to control vehicle or personnel entry into a protected area. Access control system elements include locks, electronic entry control systems, and guards.

access group A software configuration of an access control system that groups access points or authorized users for easier arrangement and maintenance of the system.

agency A division of government with a specific function offering a particular kind of assistance. In the incident command system, agencies are defined either as jurisdictional (having statutory responsibility for incident management) or as assisting or cooperating (providing resources or other assistance).

agency representative A person assigned by a primary, assisting, or cooperating agency.

annunciation A visual, audible, or other indication by a security system of a condition.

antiterrorism Preventive in nature and entails using passive and defensive measures, such as education, foreign liaison training, surveillance, and countersurveillance, designed to deter terrorist activities. The concept has two phases. The proactive phase encompasses the planning, resourcing, preventive measures, preparation, awareness education, and training that take place before a terrorist incident. The reactive phase includes the crisis management actions taken to resolve a terrorist incident.

area command (unified area command) An organization established to (a) oversee the management of multiple incidents that are each being handled by an incident command system organization or (b) oversee the management of large or multiple incidents to which several incident management teams have been assigned. Area command has the responsibility to set overall strategies and priorities, allocate critical resources according to priorities, ensure that incidents are properly managed, and ensure that objectives are met and strategies followed. Area command becomes unified area command when incidents are multijurisdictional.

areas of potential compromise Categories where losses can occur that will affect either a department's or an agency's minimum essential infrastructure and its ability to conduct core functions and activities.

business continuity plan An ongoing process supported by senior management and funded to ensure that the necessary steps are taken to identify the impact of potential losses, maintain viable recovery strategies and recovery plans, and ensure continuity services through personnel training, plan testing, and maintenance.

biological agents Microorganisms or toxins from living organisms with infectious or noninfectious properties that produce lethal or serious effects in plants and animals.

biometrics   The use of physical characteristics of the human body as a unique identification method.

blast vulnerability envelope   The geographical area in which an explosive device will cause damage to assets.

boundary penetration sensor   An interior intrusion sensor that detects attempts by individuals to penetrate or enter a building.

building hardening   Enhanced construction that reduces vulnerability to external blast and ballistic attacks.

capability maturity models   Tools for addressing software engineering and other disciplines that have an effect on software development and maintenance. Developed at Carnegie Mellon University with Department of Defense funding.

Category A diseases/agents   Highest priority agents that pose a risk to national security because they can be easily disseminated or transmitted from person to person, can result in high mortality rates and have the potential for major public health impact, might cause public panic and social disruption, and require special action for public health preparedness. The Category A agents are smallpox, anthrax, plague, botulism, tularemia, and viral hemorrhagic fevers (e.g., Ebola and Lassa viruses).

Category B diseases/agents   Second highest priority agents include those that are moderately easy to disseminate, result in moderate morbidity rates and low mortality rates, and require specific enhancements of CDC's diagnostic capacity and enhanced disease surveillance. Category B diseases are brucellosis, epsilon toxin of Clostridium perfringens, food safety threats (e.g., Salmonella species, Escherichia coli O157.H7, and Shingella), glanders, melioidosis, psittacosis, Q fever, ricin toxin, staphylococcal enterotoxin B, typhus fever, viral encephalitis (e.g., Venezuelan equine encephalitis, eastern and western encephalitis), and water safety threats (e.g., Vibrio cholerae, Cryptosporidium parvum).

Category C diseases/agents   Third highest priority agents include emerging pathogens that could be engineered for mass dissemination in the future because of the availability, ease of production and dissemination, and potential for high morbidity and mortality rates and major health impact. The Centers for Disease Control and Prevention cites Nipah virus and hantavirus as examples.

chief   The incident command system title for individuals responsible for management of functional sections, operations, planning, logistics, finance/administration, and intelligence (if established as a separate section).

choking agents   Compounds that injure an unprotected person chiefly in the respiratory tract (the nose, throat, and particularly the lungs). In extreme cases, membranes swell, lungs become filled with liquid, and death results from lack of oxygen; thus these agents "choke" an unprotected person. Choking agents include phosgene, diphosgene, and chlorine.

civil support   Department of Defense support to U.S. civil authorities for domestic emergencies, designated law enforcement, and other activities.

clear zone   An area that is clear of visual obstructions and landscape materials that could conceal a threat or perpetrator.

color infrared   Type of photographic emulsion used for remote sensing and other applications.

command and control   The exercise of authority and direction by a properly designated commander over assigned or attached forces in the accomplishment of a mission. Command and control functions are performed through an arrangement of personnel, equipment, communications, computers, facilities, and procedures employed by a commander who plans, directly coordinates, and controls forces and operations to accomplish a mission.

command staff   In an incident management organization, the command staff consists of the incident command and the special staff positions of public information officer, safety officer, liaison officer, and other positions as required, who report directly to the incident commander. They may have an assistant or assistants, as needed.

commercial off-the-shelf   Commercially available software or systems that are ready to use and which do not require significant customization.

common operating picture   A broad view of the overall situation as reflected by situation reports, aerial photography, and other information or intelligence.

competitive grant   Grant for which eligible applicants are solicited to submit concept papers. At the conclusion of the solicitation period, all received concept papers are assessed and ranked. The highest ranked applicants are then eligible for an award upon their completion of all necessary administrative requirements. Their award amount may be linked to their ranking.

congregate care center   A public or private facility predesignated and managed by the American Red Cross during an emergency where evacuated or displaced persons are housed and fed.

consequence management   Measures to alleviate the damage, loss, hardship, or suffering caused by emergencies. It includes measures to restore essential government service, protect public health and safety, and provide emergency relief to affected governments, businesses, and individuals. Per HSPD-5, crises management and consequence management are merged into a single integrated function called domestic incident management.

continuity of government   Planning that ensures the continuity of essential functions in any state security emergency by providing for succession to office and emergency delegation of authority in accordance with applicable law, safekeeping of essential resources, facilities, and records, and establishment of emergency operating capabilities.

continuity of operations   Efforts taken within an entity (i.e., agency, company, association, organization, or business) to ensure continuance of minimum essential functions across a wide range of potential emergencies, including localized acts of nature, accidents, and technology- or attack-related emergencies.

continuously operating reference station   A Global Positioning System correction facility that captures and archives satellite positions and vectors that can be used for establishing precise longitude, latitude, and elevation positions from a high-end GPS receiver.

crisis management   Measures to resolve a hostile situation, investigate, and prepare a criminal case for prosecution under federal law. Per HSPD-5, crisis management and consequence management are merged into a single integrated function referred to as domestic incident management.

critical agents   The biological and chemical agents likely to be used in weapons of mass destruction and other bioterrorist attacks.

critical information   Specific facts about friendly intentions, capabilities, and activities vitally needed by adversaries to plan and act effectively to guarantee failure or unacceptable consequences.

critical infrastructure   Those systems and assets—both physical and cyber—so vital to the states, localities, and the nation that their incapacity or destruction would have a debilitating impact on national, state, and local security, economic security; or public health and safety.

cyber infrastructure   Within our critical infrastructure sectors (agriculture and food, water, health care and public health, emergency services, government facilities, defense manufacturing capability, information and telecommunications, energy, transportation, banking and finance, chemical and hazardous materials, postal and shipping) those cyber-related (continuum of computer networks) information technology systems and assets (e.g., interconnected computer networks, automated control systems, information systems, servers, router switches, and fiber-optic cables) that allow our critical infrastructure systems to function.

cyberterrorism   A criminal act perpetrated by using computers and telecommunications, resulting in violence, destruction, and/or disruption of services. Such acts create fear by causing confusion and uncertainty within a given population with the goal of influencing a government or population to conform to a particular political, social, or ideological agenda.

damage assessment   The process used to appraise or determine the number of injuries and deaths, damage to public and private property, and the status of key facilities and services (e.g., hospitals and other health care facilities, fire and police stations, communications networks, water and sanitation systems, utilities, and transportation networks) resulting from a human-made or natural disaster.

data   Unprocessed, unanalyzed raw observations and facts.

deconfliction   A process designed to prevent, coordinate, or preclude duplicate investigations by two separate individuals or agencies. Deconfliction systems are generally computer-based, some with GIS mapping capability linked to a database.

digital line graph   A U.S. Geological Survey data product for mapping vectors in the agency's standard format, supporting 1:24,000-scale and other maps.

digital orthophoto quadrangle   An aerial photograph that has undergone some geometric correction to approximate the geometry and extent of a U.S. Geological Survey 1:24,000-scale quadrangle map sheet. Digital orthophoto quadrangle is a quarter quad (one fourth of the quad area).

digital terrain elevation data   A format and data comprising a uniform matrix of terrain elevation values, which provides basic quantitative data for systems and applications that require terrain elevation, slope, and/or surface roughness information.

disaster field office   The office established in or near the designated area of a presidentially declared major disaster to support federal and state response and recovery operations.

disaster or emergency declaration   A declaration by the president that authorizes supplemental federal assistance under the Robert T. Stafford Disaster Relief and Emergency Assistance Act. The declaration is in response to a governor's request and may cover a range of response, recovery, and mitigation assistance for state and local governments, eligible private nonprofit organizations, and individuals.

disaster recovery center   Places established in the area of a presidentially declared major disaster, as soon as practicable, to provide victims the opportunity to apply in person for assistance or obtain information relating to that assistance.

discretionary grant   Federal grant funds distributed to states, units of local government, or private organizations at the discretion of the agency administering the funds. Most discretionary grants are competitive and usually have limited funds available and a large number of potential recipients.

division   The partition of an incident into geographical areas of operation. Divisions are established when the number of resources exceeds the manageable span of control of the operations chief.

domain awareness   The integration of information, intelligence, sensors, and surveillance capabilities over an area (domain) to provide effective knowledge of all activities and elements in the domain that threaten the safety, security, or environment of the United States or its citizens. In the lexicon of national homeland security strategic objectives, domain awareness falls under the prevention objective and is primarily attributed to maritime environments (ports, rivers, oceans).

emergency   Any incident, human-caused or natural, that requires responsive action to protect life or property. Under the Stafford Act, an emergency means any occasion or instance for which, in the determination of the president, federal assistance is needed to supplement state and local efforts and capabilities to save lives and protect property and public health and safety or to lessen or avert the threat of a catastrophe in any part of the United States.

emergency environmental health services   Services required to correct or improve damaging environmental health effects on humans, including inspection for food contamination, inspection for water contamination, and vector control; providing for sewage and solid waste inspection and disposal; cleanup and disposal of hazardous materials; and sanitation inspection for emergency shelter facilities.

emergency management   The process by which the states and the nation prepare for emergencies and disasters, mitigate their effects, and respond to and recover from them.

emergency operations center   A secure location to determine situational status, coordinate actions, and make critical decisions during emergency and disaster situations.

emergency operations plan   A planning document that (a) assigns responsibility to organizations and individuals for implementing specific actions at projected times and places in an emergency that exceeds the capability or routine responsibility of any one agency; (b) sets forth lines of authority and organizational relationships and shows how all actions will be coordinated; (c) identifies personnel, equipment, facilities, supplies, and other resources available for use during response and recovery operations; and (d) identifies steps to address mitigation issues during response and recovery activities.

emergency planning zone   Areas around a facility for which planning is needed to ensure prompt and effective actions are taken to protect the health and safety of the public if an accident or disaster occurs. In the Radiological Emergency Preparedness Program, the plume exposure pathway (ten-mile emergency planning zone) is a circular geographic zone (with a ten-mile radius centered at the nuclear power plant) for which plans are developed to protect the public against exposure to radiation emanating from a radioactive plume caused as a result of an accident at the nuclear power plant.

emergency public information   Information disseminated primarily in anticipation of an emergency or during an emergency. In addition to providing situational information to the public, it also frequently provides directive actions required to be taken by the general public.

emergency response coordinator   Person authorized to direct implementation of an agency's emergency response plan.

emergency response provider   Includes federal, state, local, and tribal emergency public safety, law enforcement, emergency response, emergency medical (including hospital emergency facilities) and related personnel, agencies, and authorities.

emergency services   A critical infrastructure characterized by medical, police, fire, and rescue systems and personnel called upon when an individual or community is responding to emergencies. These services are typically provided at the local level. In addition, state and federal response plans define emergency support functions that can assist in response and recovery.

emergency support function   The functional approach that groups the types of assistance that a state is most likely to need (e.g., mass care, health and medical services) as well as the kinds of federal operations support necessary to sustain state response actions (e.g., transportation, communications). Emergency support functions are expected to support one another in carrying out their respective missions.

essential elements of information   Key questions likely to be asked by adversary officials and intelligence systems about our specific friendly intentions, capabilities, and activities, so they can obtain answers critical to their operational effectiveness. Also called essential elements of friendly information.

extensible markup language   A variety of HTML. Extensible Markup Language (XML) is a simple, very flexible text format derived from SGML (ISO 8879). Originally designed to meet the challenges of large-scale electronic publishing, XML is also playing an increasingly important role in the exchange of a wide variety of data on the Web and elsewhere.

extract, transform, and load   Steps used to migrate (import and export) geographic data from one format or source to another while retaining most or all of the significance of the data.

farmgate   The value of production for all agricultural products.

field assessment team   A small team of preidentified technical experts that conduct an assessment of response needs immediately following a disaster.

fire service   Individuals, who on a full-time, volunteer, or part-time basis provide life safety services, including fire suppression, rescue, arson investigation, public education, and prevention.

first responders   Local police, firefighters, and emergency medical professionals who are the first to arrive at the scene of an emergency or terrorist attack.

focus areas   Categories of emergency preparedness activities states must address in their Cooperative Agreements for Public Health Preparedness and Response for Bioterrorism. Focus areas cover preparedness planning and readiness assessment; surveillance and epidemiology capacity; laboratory capacity for biological and chemical agents; health alert network/communications and information technology; communicating health risk and health information dissemination; and education and training.

function   Refers to the five major activities in an incident command system: command, operations, planning, logistics, and finance/administration. A sixth function, intelligence, may be established, if required, to meet incident management needs.

fusion center   An organized structure to coalesce data and information for the purpose of analyzing, linking, and disseminating intelligence. A model process likely extracts unstructured and structured data, and fuses structured data. Fused data is analyzed to generate intelligence products and summaries for tactical, operational, and strategic commanders. Types of analysis typically conducted in a fusion center include association charting, temporal charting, spatial charting, link analysis, financial analysis, content analysis, and correlation analysis.

general staff   A group of incident management personnel organized according to function and who report to the incident commander. The general staff normally consists of the operations section chief, planning section chief, logistics section chief, and finance administration section chief.

geographic markup language   This is extensible markup language encoding for the modeling, transport, and storage of geographic information.

governmental administrative   Elected and appointed officials responsible for public administration of community health and welfare during a weapons of mass destruction terrorism incident.

group   Established to divide the incident management structure into functional areas of operation. Groups are composed of resources assembled to perform a special function not necessarily within a single geographic division. Groups, when activated, are located between branches and resources in the operations section.

hazard   Something potentially dangerous or harmful, often the root cause of an unwanted outcome.

hazard mitigation   Any action taken to reduce or eliminate the long-term risk to human life and property from hazards. The term is sometimes used in a stricter sense to mean cost-effective measures to reduce the potential for damage to a facility or facilities from a disaster event.

hazardous materials personnel   Individuals, who on a full-time, volunteer, or part-time basis identify, characterize, provide risk assessment, and mitigate or control the release of a hazardous substance or potentially hazardous substance.

health alerts   Urgent messages from the Centers for Disease Control and Prevention to health officials requiring immediate action or attention. The agency also issues health advisories containing less urgent information about a specific health incident or response that may or may not require immediate action as well as health updates, which do not require action.

high-hazard areas   Geographic locations that, for planning purposes, have been determined through historical experience and vulnerability analysis to be likely to experience the effects of a specific hazard (e.g., hurricane, earthquake, hazardous materials accident), resulting in vast property damage and loss of life.

high-risk target   Any material resource or facility that, because of mission sensitivity, ease of access, isolation, and symbolic value, may be an especially attractive or accessible terrorist target.

homeland defense   The protection of U.S. territory, sovereignty, domestic population, and critical infrastructure against external threats and aggression.

homeland security   (1) A concerted national effort to prevent terrorist attacks within the United States, reduce America's vulnerability to terrorism, and minimize the damage and recover from attacks that do occur. (2) The preparation for, prevention of, deterrence of, preemption of, defense against, and response to threats and aggressions directed toward U.S. territory, sovereignty, domestic populations, and infrastructure as well as crisis management, consequence management, and other domestic civil support.

hotwash   An after-action review of events or training that discusses what went right, what went wrong, and what to do differently next time.

immediate response zone   A circular zone ranging from six to nine miles from the potential chemical event source, depending on the stockpile location on post. Emergency response plans developed for this zone must provide for the most rapid and effective protective actions possible, because the zone will have the highest concentration of agent and the least amount of warning time.

impact analysis   A management-level analysis that identifies the impacts of losing the entity's resources. The analysis measures the effect of resource loss and escalating losses over time to provide the entity with reliable data upon which to base decisions about hazard mitigation and continuity planning.

incapacitating agents   An agent that produces temporary physiological or mental effects via action on the central nervous system. Effects may persist for hours or days, and victims usually do not require medical treatment; however, such treatment does speed recovery.

incident   An occurrence or event, natural or human-caused, that requires an emergency response to protect life or property. Incidents can include major disasters, emergencies, terrorist attacks, terrorist threats, wild land and urban fires, floods, hazardous materials spills, nuclear accidents, aircraft accidents, earthquakes, hurricanes, tornadoes, tropical storms, war-related disasters, and public health and medical emergencies.

incident action plan   An oral or written plan containing general objectives reflecting the overall strategy for managing an incident. It may include the identification of operational resources and assignments. It may also include attachments that provide direction and important information for management of the incident during one or more operational periods. Contains objectives reflecting the overall incident strategy, specific tactical actions, and supporting information for the next operational period. When written, the plan may have a number of forms as attachments (e.g., traffic plan, safety plan, communications plan, and maps).

incident commander   The individual responsible for all incident activities, including the development of strategies and tactics and the ordering and release of resources. The incident commander has overall authority and responsibility for conducting incident operations and is responsible for the management of all incident operations at the incident site.

incident command post   The field location at which the primary tactical-level, on-scene incident command functions are performed. The incident command post may be colocated with the incident base or other incident facilities and is normally identified by a green rotating or flashing light.

incident command system   A model for disaster response that calls for the use of common terminology, modular organization, integrated communications, unified command structure, action planning, manageable span-of-control, predesignated facilities, and comprehensive resource management.

incident management team   The incident commander and appropriate command and general staff personnel assigned to an incident.

incident objectives   Statements of guidance and direction necessary for selecting appropriate strategies and the tactical direction of resources. Incident objectives are based on realistic expectations of what can be accomplished when all allocated resources have been effectively deployed. Incident objectives must be achievable and measurable, yet flexible enough to allow strategic and tactical alternatives.

infrastructure   The framework of interdependent networks and systems comprising identifiable industries, institutions (including people and procedures), and distribution capabilities that provide a reliable flow of products and services essential to the defense and economic security of the United States and the smooth functioning of governments at all levels of society as a whole.

ingestion pathway (fifty-mile emergency planning zone)   A circular geographic zone (with a fifty-mile radius centered at the nuclear power plant) for which plans are developed to protect the public from the ingestion of water or food contaminated as a result of a nuclear power plant accident.

initial response   Resources initially committed to an incident.

interoperability   The ability of systems or communications to work together.

joint field office   Federal activities at a local incident site will be integrated during domestic incidents to better facilitate coordination among federal, state, and local authorities. The joint field office is expected to incorporate existing entities such as the joint operations center, the disaster field office, and other federal offices and teams that provide support on scene.

joint information center   A central point of contact for all news media near the scene of a large-scale disaster. The center is staffed by public information officials who represent all participating federal, state, and local agencies to provide information to the media in a coordinated and consistent manner.

joint information system   Integrates incident information and public affairs into a cohesive organization designed to provide consistent, coordinated, and timely information during crisis or incident operations. The mission of the joint information system is to provide a structure and system for developing and delivering coordinated interagency messages; developing, recommending, and executing public information plans and strategies on behalf of the incident commander; advising the incident commander concerning public affairs issues that could affect a response effort; and controlling rumors and inaccurate information that could undermine public confidence in the emergency response effort.

jurisdiction   A range or sphere of authority. Public agencies have jurisdiction at an incident related to their legal responsibilities and authority. Jurisdictional authority at an incident can be political or geographic (e.g., city, county, tribal, state, or federal boundary lines) or functional (e.g., law enforcement, public health).

laboratory levels (A, B, C, D)   A system for classifying laboratories by their capabilities.

A. Routine clinical testing. Includes independent clinical labs and those at universities and community hospitals.

B. More specialized capabilities. Includes many state and local public health laboratories.

C. More sophisticated public health labs and reference labs such as those run by Centers for Disease Control and Prevention.

D. Possessing sophisticated containment equipment and expertise to deal with the most dangerous, virulent pathogens and including only Department of Defense and Centers for Disease Control and Prevention labs, the FBI, and the U.S. Army Medical Research Institute of Infectious Diseases.

local government   A county, municipality, city, town, township, local public authority, school district, special district, intrastate district, council of governments (regardless of whether the council of governments is incorporated as a nonprofit corporation under state law), regional or interstate government entity, or agency or instrument of a local government; an Indian tribe or authorized tribal organization or, in Alaska, a Native village or Alaska Regional Native Corporation; or a rural community, unincorporated town or village, or other public entity. See Section 2 (10), Homeland Security Act of 2002, Pub. L. 107-296, 116 Stat. 2135 (2002).

logistics   Providing resources and other services to support incident management.

major disaster   As defined under the Stafford Act, any natural catastrophe (including hurricane, tornado, storm, high water, wind-driven water, tidal wave, tsunami, earthquake, volcanic eruption, landslide, mudslide, snowstorm, or drought) or, regardless of cause, any fire, flood, or explosion in any part of the United States, which in the determination of the president causes damage of sufficient severity and magnitude to warrant major disaster assistance under this act to supplement the efforts and available resources of states, tribes, local governments, and disaster relief organizations in alleviating the damage, loss, hardship, or suffering caused thereby.

management by objective   A management approach that involves a four-step process for achieving the incident goal. The management by objective approach includes the following: establishing overarching objectives; developing and issuing assignments, plans, procedures, and protocols; establishing specific, measurable objectives for various incident management functional activities and directing efforts to fulfill them in support of defined strategic objectives; and documenting results to measure performance and facilitate corrective action.

mass care   Actions taken to protect evacuees and other disaster victims from the effects of the disaster. Activities include providing temporary shelter, food, medical care, clothing, and other essential life support needs to those people who have been displaced from their homes because of a disaster or threatened disaster.

mass notification   Capability to provide real-time information to all building occupants or personnel in the immediate vicinity of a building during emergency situations.

mitigation   The activities designed to reduce or eliminate risks to persons or property or to lessen the actual or potential effects or consequences of an incident. Mitigation measures may be implemented prior to, during, or after an incident. Mitigation measures are often informed by lessons learned from prior incidents. Mitigation involves ongoing actions to reduce exposure to, probability of, or potential loss from hazards. Measures may include zoning and building codes, floodplain buyouts, and analysis of hazard-related data to determine where it is safe to build or locate temporary facilities. Mitigation can include efforts to educate governments, businesses, and the public on measures they can take to reduce loss and injury.

multiagency coordination systems   These systems provide the architecture to support coordination for incident prioritization, critical resource allocation, communications systems integration, and information coordination. The components of multiagency coordination systems include facilities, equipment, emergency operation centers, specific multiagency coordination entities, personnel, procedures, and communications. These systems assist agencies and organizations to fully integrate the subsystems of the National Incident Management System.

multijurisdictional incident   An incident requiring action from multiple agencies that each have jurisdiction to manage certain aspects of an incident. In an incident command system, these incidents will be managed under unified command.

mutual-aid agreement   Written agreement between agencies and/or jurisdictions that they will assist one another, on request, by furnishing personnel, equipment, or expertise in a specified manner.

National Map   A U.S. Geological Survey data site for dissemination of map data. It provides public access to high-quality, geospatial data and information from multiple partners to help support decision making by resource managers and the public.

national security emergency   Any occurrence, including natural disaster, military attack, technological emergency, or other emergency, that seriously degrades or threatens the national security of the United States (Executive Order 12656).

nerve agent   Organophosphate ester derivatives of phosphoric acid. Potent inhibitors of the enzyme acetylcholinesterase (AChE), causing a disruption in normal neurologic function. Symptoms appear rapidly with death occurring as rapidly as several minutes. Nerve agents are generally divided into G-series agents and V-series agents and include tabun (GA), sarin (GB), soman (GD), and 0-Ethyl S-(2-Diisopropylaminoethyl) Methylphosphonothioate (VX).

nongovernmental organization   An entity with an association based on interests of its members, individuals, or institutions and is not created by, but may work cooperatively with, government. Such organizations serve a public purpose, not a private benefit. Examples of nongovernmental organizations include faith-based charity organizations and the American Red Cross.

nonpersistent agent   An agent that, upon release, loses its ability to cause casualties after ten to fifteen minutes. It has a high evaporation rate, is lighter than air, and will disperse rapidly. A nonpersistent agent is considered to be a short-term hazard; however, in small, unventilated areas, the agent will be more persistent.

on-scene commander   A term used to designate the FBI person who provides leadership and direction to the federal crisis management response. The FBI on-scene commander may or may not be the regional Special Agent in Charge.

open systems architecture   A term borrowed from the information technology industry to claim that systems are capable of interfacing with other systems from any vendor that also uses open system architecture. The opposite would be a proprietary system.

operational period   The time scheduled for executing a given set of operation actions, as specified in the Incident Action Plan. Operational periods can be of various lengths, although usually not more than twenty-four hours.

performance measure   A specific measurable result for each goal that indicates successful achievement.

personnel accountability   The ability to account for the location and welfare of incident personnel. It is accomplished when supervisors ensure that incident command system principles and processes are functional and that personnel are working within established incident management guidelines.

potential threat element   Any group or individual in which there are allegations or information indicating a possibility of the unlawful use of force or violence, specifically, the use of a weapon of mass destruction against persons or property to intimidate or coerce a government, the civilian population, or any segment thereof, in furtherance of a specific motivation or goal, possibly political or social in nature. This definition provides sufficient predicate for the FBI to initiate an investigation.

prearrival processing system   A U.S. Customs automated commercial system border cargo release mechanism that uses bar code technology to expedite the release of commercial shipments while processing each shipment through Border Cargo Selectivity and the Automated Targeting System.

precautionary zone   The outermost portion of the emergency planning zone for the Chemical Stockpile Emergency Preparedness Program, extending from the protective action zone outer boundary to a distance where the risk of adverse impacts on humans is negligible. Because of the increased warning and response time available for implementation of response actions in the precautionary zone, detailed local emergency planning is not required, although consequence management planning may be appropriate.

preempt   Acting emergency to eliminate an opponent's ability to take a specific action, preventing them with efforts in surveillance, detection, intelligence gathering/sharing, cooperation, early warning, and effective command and control.

preparedness   The range of deliberate, critical tasks and activities necessary to build, sustain, and improve the operational capability to prevent, protect against, respond to, and recover from domestic incidents. Preparedness is a continuous process. Preparedness involves efforts at all levels of government and between government and private sector and nongovernmental organizations to identify threats, determine vulnerabilities, and identify required resources.

prevention   Actions to avoid an incident or to intervene to stop an incident from occurring. Prevention involves actions to protect lives and property. It involves applying intelligence and other information to a range of activities that may include such countermeasures as deterrence operations; heightened inspections; improved surveillance and security operations; investigations to determine the full nature and source of the threat; public health and agricultural surveillance and testing processes; immunizations, isolation, or quarantine; and, as appropriate, specific law enforcement operations aimed at deterring, preempting, interdicting, or disrupting illegal activity and apprehending potential perpetrators and bringing them to justice.

principal federal official   The secretary may designate a principal federal official during a domestic incident to serve as the representative of the Department of Homeland Security locally during an incident. This official oversees and coordinates federal incident activities and works with local authorities to determine requirements and provide timely federal assistance.

private sector   Organizations and entities that are not part of any governmental structure. It includes for-profit and not-for-profit organizations; formal and informal structures; commerce and industry; and private, voluntary organizations.

Project BioShield   A comprehensive effort to develop and make available modern, effective drugs and vaccines to protect against attack by biological and chemical weapons or other dangerous pathogens. Project BioShield will ensure that resources are available to pay for next-generation medical countermeasures. Project BioShield will allow the government to buy improved vaccines or drugs for smallpox, anthrax, and botulinum toxin.

protect   Protection consists of six groups of activities: hardening of positions; protecting personnel; assuming mission-oriented, protective posture; hardening of positions (infrastructure); protecting people; using physical defense measures; and reacting to an attack (JCS Pub. 1-02).

protective action zone   An area that extends beyond the immediate response zone to approximately sixteen to fifty kilometers (ten to thirty miles) from the stockpile location. The protective action zone is that area where public protective actions may still be necessary in case of an accidental release of chemical agent but where the available warning and response time is such that most people could evacuate. However, other responses (e.g., sheltering) may be appropriate for institutions and special populations that could not evacuate within the available time

public health regions   Local health jurisdictions are organized into nine regions. Each region will develop a plan for resource sharing and coordinated emergency response that will align to the state emergency management plan and will include hospitals, emergency medical services, law enforcement, and fire protection districts.

publications management   A publications management subsystem that includes materials development, publication control, publication supply, and distribution. The development and distribution of National Incident Management System materials are managed through this subsystem. Consistent documentation is critical to success, because it ensures that all responders are familiar with the documentation used in a particular incident regardless of the location or the responding agencies involved.

push package   A delivery of medical supplies and pharmaceuticals sent from the National Pharmaceutical Stockpile to a state undergoing an emergency within twelve hours of federal approval of a request by the state's governor.

qualification and certification   A subsystem that provides recommended qualification and certification standards for emergency responder and incident management personnel. It also allows the development of minimum standards for resources expected to have an interstate application. Standards typically include training, currency, experience, and physical and medical fitness.

radiological dispersal devices   A conventional explosive device incorporating radioactive material(s) sometimes referred to as a "dirty bomb."

rapid response information system   A system of databases and links to Internet sites providing information to federal, state, and local emergency officials on federal capabilities and assistance available to respond to consequences of a weapons of mass destruction/ terrorism incident. This information is available to designated officials in each state, the ten FEMA regions, and key federal agencies via a protected Internet site and indirectly to the Intranet site through their respective state counterparts. It can be used as a reference guide, training aid, and an overall planning and response resource for weapons of mass destruction/terrorism incidents.

recovery   The development, coordination, and execution of service- and site-restoration plans; the reconstitution of government operations and services; individual, private sector, nongovernmental, and public assistance programs to provide housing and promote restoration; long-term care and treatment of affected persons; additional measures for social, political, environmental, and economic restoration; evaluation of the incident to identify lessons learned; post-incident reporting; and development of initiatives to mitigate the effects of future incidents.

relational database management system   This is a type of database management system (DBMS) that stores data in the form of related tables.

resource management   Efficient incident management requires a system for identifying available resources at all jurisdictional levels to enable timely and unimpeded access to resources needed to prepare for, respond to, or recover from an incident. Resource management under National Incident Management System includes mutual-aid agreements; the use of special federal, state, local, and tribal teams; and resource mobilization protocols.

resources   Personnel and major items of equipment, supplies, and facilities available or potentially available for assignment to incident operations and for which status is maintained. Resources are described by kind and type and may be used in operational support or supervisory capacities at an incident or at an emergency operations center.

response   Activities that address the short-term, direct effects of an incident. Response includes immediate actions to save lives, protect property, and meet basic human needs. Response also includes the execution of emergency operations plans and mitigation activities designed to limit the loss of life, personal injury, property damage, and other unfavorable outcomes. As indicated by the situation, response activities include applying intelligence and other information to lessen the effects or consequences of an incident; increased security operations; continuing investigations into the nature and source of the threat; ongoing public health and agricultural surveillance and testing processes; immunizations, isolation, or quarantine; and specific law-enforcement operations aimed at preempting, interdicting, or disrupting illegal activity and apprehending actual perpetrators and bringing them to justice.

safety officer   A member of the command staff responsible for monitoring and assessing safety hazards or unsafe situations and for developing measures for ensuring personnel safety.

satellite-based technology   Communication systems that are not prone to the same outages as terrestrially based systems.

security management system database   In a security management system, a database transferred to various nodes or panels throughout the system for faster data processing and protection against communications-link downtime.

sensitive but unclassified   A type of restriction on the dissemination of data or documents not subject to rigorous classification but may still contain sensitive information.

shelter in place   To stay where one is in the event of an emergency. Requires that people stay inside a building away from windows. All windows and air intake systems should be closed. Wet towels or tape may be used to seal cracks. If there is a danger of explosion, windows should be covered.

software-level integration   An integration strategy that uses software to interface systems. An example of this is digital video displayed in the same computer application window and linked to events of a security management system.

span of control   The number of individuals for which a supervisor is responsible, usually expressed as the ratio of supervisors to individuals. (Under National Incident Management System, an appropriate span of control is between 1.3 and 1.7.)

spatial database engine   ESRI technology that provides an enterprise-wide repository for spatial and attribute data within a relational database management system

staging area   Location established where resources can be placed while awaiting a tactical assignment. The operations section manages staging areas.

strategic elements   Strategic elements of incident management are characterized by continual long-term, high-level planning by organizations headed by elected or other senior officials. These elements involve the adoption of long-range goals and objectives, the setting of priorities, the establishment of budgets and other fiscal decisions, policy development, and the application of measures of performance or effectiveness.

strategic planning   The systematic identification of opportunities and threats that lie in the future environment, both external and internal, which, in combination with other relevant data, such as threats, vulnerabilities, and risks, provides a basis to make better current decisions to pursue opportunities and to avoid threats. It is an orderly process that sets basic objectives and goals to be achieved and strategies to reach those goals and objectives with supporting action plans to ensure that strategies are properly implemented.

strike team   A set number of resources of the same kind and type that have an established minimum number of personnel.

supplanting   Deliberately reducing state or local funds because of the existence of federal funds.

supporting technologies   Any technology that may be used to support National Incident Management System is included in this subsystem. These technologies include orthophoto mapping, remote automatic weather stations, infrared technology, and communications, among others.

surge capacity   Ability of institutions, such as clinics, hospitals, or public health laboratories, to sharply increase demand for their services during an emergency.

surveillance   Looking at the background level to check for the presence of disease. An example would be a case in which the health department contracts with a farmer to raise chickens then tests the blood of the chickens for the presence of disease.

system software   Controls that limit and monitor access to the powerful programs and sensitive files controlling the computer hardware and secure applications supported by the system.

target capabilities list   Mandated by HSPD-8, will specify universal task lists that responders will have to manage for homeland security scenarios.

terrorism   The unlawful use of force or violence against persons or property to intimidate or coerce a government, the civilian population, or any segment thereof, in furtherance of political or social objectives. Domestic terrorism involves groups or individuals who are based and operate entirely within the United States and U.S. territories without foreign direction and whose acts are directed at elements of the U.S. government or population.

terrorist early warning   A group of experts working in a common secure area that receives data feeds from intelligence sources.

terrorist incident   The FBI defines a terrorist incident as a violent act or an act dangerous to human life, in violation of the criminal laws of the United States or of any state, to intimidate or coerce a government, the civilian population, or any segment thereof in furtherance of political or social objectives.

threat   An indication of possible violence, harm, or danger.

threat analysis   A continual process of compiling and examining all available information concerning potential threats and human-caused hazards. A common method to evaluate terrorist groups is to review the factors of existence, capability, intentions, history, and targeting.

triba   Any Indian tribe, band, nation, or other organized group or community, including any Alaskan Native Village as defined in or established pursuant to the Alaskan Native Claims Settlement Act (85 Stat. 688) [43 U.S.C.A. and 1601 et seq.], that is recognized as eligible for the special programs and services provided by the United States to Indians because of their status as Indians.

type   A classification of resources in incident command systems that refers to capability. Type 1 is usually considered to be more capable than types 2, 3, or 4, respectively, because of size; power; capacity; or in the case of incident management teams, experience and qualifications.

unified area command   A unified area command is established when incidents under an area command are multijurisdictional.

unified command   An application of the incident command system used when there is more than one agency with incident jurisdiction or when incidents cross political jurisdictions. Agencies work together through the designated members of unified command, often the senior persons from agencies or disciplines participating in unified command, to establish a common set of objectives and strategies and a single incident action plan.

unity of command   The concept by which each person within an organization reports to one and only one designated person. The purpose of unity of command is to ensure unity of effort under one responsible commander for every objective.

universal description, discovery, and integration   This is a protocol that creates a standard interoperable platform that enables companies and applications to quickly, easily, and dynamically find and use Web services over the Internet.

universal task list   Created by the Department of Homeland Security, the universal task list defines the essential tasks to be performed by federal, state, and local governments and the private sector to prevent, respond to, and recover from a range of threats from terrorists, natural disasters, and other emergencies.

vulnerability   (1) The susceptibility of a nation or military force to any action by any means through which its war potential or combat effectiveness may be reduced or its will to fight diminished. (2) The characteristics of a system that cause it to suffer a definite degradation (incapacity to perform the designated mission) as a result of having been subjected to a certain level of effects in an unnatural (man-made) hostile environment. (3) In information operations, a weakness in information systems security design, procedures, implementation, or internal controls that could be exploited to gain unauthorized access to information or an information system (JCS Pub. 1-02).

vulnerability assessment   The vulnerability assessment provides a measure to indicate the relative likelihood that a particular facility or incident within the jurisdiction may become the target of a terrorist attack. The factors considered include measures of attractiveness and impact.

watch-out situations   In fire management and fire service, watch-out situations are indicators or trigger points that remind firefighters to reanalyze or to reevaluate their suppression strategies and tactics. The watch-out situations in the fire service are more specific and cautionary than the "Ten Standard Fire Orders." In antiterrorism, the term is used as a metaphor for those observations that can alert trained personnel—not only firefighters but also law enforcement, public works, private security, or anyone—to be more cautious, more observant, and more likely to report the unusual behavior or activity to the appropriate authorities.

weapons of mass destruction   Any explosive, incendiary, or poison gas; bomb, grenade, rocket having a propellant charge of more than four ounces; missile having an explosive incendiary charge of more than 0.25 ounce; mine or device similar to the above; weapon involving a disease organism; or weapon designed to release radiation or radioactivity at a level dangerous to human life. (Source: 18 USC 2332a as referenced in 18 USC 921.)

Web services description language   An extensible markup language format for describing network services as a set of endpoints operating on messages containing either document-oriented or procedure-oriented information.

zoonotic   Of or relating to zoonosis. An animal disease that can be transmitted to humans (e.g., Ebola, Lyme disease, anthrax, rabbit fever, rabies, and swamp fever).

Sources

Compendium of Federal Terrorism Training for State and Local Audiences.
http://www.dola.colorado.gov/dem/publications/FEMA_Training_Compendium.pdf.

NIMS Glossary of Key Terms.
http://www.nimsonline.com/nims_3_04/glossary_of_key_terms.htm.

FEMA. http://training.fema.gov/EMIWeb/IS/is14/glossary.htm.

Homeland Security Glossary of Terms.
http://www.dhs.gov/xlibrary/assets/foia/US-VISIT_RFIDattachB.pdf.

Joint Chiefs of Staff Publications. http://www.dtic.mil/doctrine/s_index.html.

National Mutual Aid & Resource Management Initiative Glossary of Terms and Definitions.
http://www.fema.gov/pdf/emergency/nims/507_Mutual_Aid_Glossary.pdf.

Legal Information Institute, U.S. Code Library. http://www.law.cornell.edu/uscode/.

Related titles from ESRI Press

Beyond Maps: GIS and Decision Making in Local Government
ISBN: 1-879102-79-X

Confronting Catastrophe: A GIS Handbook
ISBN: 1-58948-040-6

Disaster Response: GIS for Public Safety
ISBN: 1-879102-88-9

GIS in Public Policy
ISBN: 1-879102-66-8

ESRI Press publishes books about the science, application, and technology of GIS. Ask for these titles at your local bookstore or order by calling 1-800-447-9778. You can also read book descriptions, read reviews, and shop online at www.esri.com/esripress. Outside the United States, contact your local ESRI distributor.